Techniques for Predicting Metal Mining Influenced Water

Techniques for Predicting Metal Mining Influenced Water

Editor

Sumit Gupta

scitus
academics

Techniques for Predicting Metal Mining Influenced Water

Edited by **Sumit Gupta**

Printed in 2017

ISBN: 978-1-68117-496-9

Library of Congress Control Number: 2015936673

© 2016 by
SCITUS Academics LLC,
616, Corporate Way, Suite 2, 4766,
Valley Cottage, NY 10989

www.scitusacademics.com

Contents

Preface

Techniques for Predicting Metal Mining Influenced Water is a must-read for planners, regulators, consultants, land managers, researchers, students, stakeholders, and others concerned about mining influenced water. Identifying potential mine wastes and their characteristics, and predicting their drainage quality are critical aspects of mine site design, operations, and closure planning. Failure to effectively conduct these evaluations for a mine site can result in environmental compliance issues that may create long-term financial liabilities. Basics of Metal Mining Influenced Water serves as an introduction to a series of six handbooks on technologies for management of metal mine and metallurgical process drainage. The other five handbooks are Mitigation of Metal Mining Influenced Water; Mine Pit Lakes: Characteristics, Predictive Modeling, and Sustainability; Geochemical Modeling for Mine Site Characterization and Remediation; Techniques for Predicting Metal Mining Influenced Water; and Sampling and Monitoring for the Mine Life Cycle.

Editor

Measured Copper Toxicity to Cnesterodon decemmaculatus(Pisces: Poeciliidae) and Predicted by Biotic Ligand Model in Pilcomayo River Water: A Step for a Cross-Fish-Species Extrapolation

María Victoria Casares[1], Laura I. de Cabo[1], Rafael S. Seoane[2, 3], Oscar E. Natale[2], Milagros Castro Ríos[2], Cristian Weigandt[4], and Alicia F. de Iorio[4]

ardino Rivadavia National Museum of Natural History, Avenida Angel Gallardo 470, C1405DJR Buenos Aires, Argentina

[2]National Water Institute, Autopista Ezeiza-Cañuelas, Tramo Jorge Newbery km 1.62 (1802), Ezeiza, C1004AA1 Buenos Aires, Argentina

[3]Faculty of Engineering, University of Buenos Aires, Avenida Las Heras 2214, C1127AAR Buenos Aires, Argentina

[4]Faculty of Agronomy, University of Buenos Aires, Avenida San Martín 4453, C1417DSE Buenos Aires, Argentina

ABSTRACT

In order to determine copper toxicity (LC50) to a local species (Cnesterodon decemmaculatus) in the South American Pilcomayo River water and evaluate a cross-fish-species extrapolation of Biotic Ligand Model, a 96 h acute copper toxicity test was performed. The dissolved copper concentrations tested were 0.05, 0.19, 0.39, 0.61, 0.73, 1.01, and 1.42 mg Cu L^{-1}. The 96 h Cu LC50 calculated was 0.655 mg L^{-1} (0.823$-$0.488). 96-h Cu LC50 predicted by BLM for Pimephales promelas was 0.722 mg L^{-1}. Analysis of the inter-seasonal variation of the main water quality parameters indicates that a higher protective effect of calcium, magnesium, sodium, sulphate, and chloride is expected during the dry season. The very high load of total suspended solids in this river might be a key factor in determining copper distribution between solid and solution phases. A cross-fish-species extrapolation of copper BLM is valid within the water quality parameters and experimental conditions of this toxicity test.

INTRODUCTION

The number of large-scale mining operations has been increasing greatly in Argentina during the last decade. It has resulted in social and environmental conflicts of diverse scale [1]. Some river basins are seriously polluted by heavy metals released by present and ancient mining activities [2]. Furthermore, occasional accidents have aggravated this situation by suddenly introducing substantial amounts of heavy metals into aquatic environments, which might be accompanied by changes in water pH, depending on the type of the mining effluent in question. Some tributaries to the upper Pilcomayo River, in Bolivia, drain a large conical peak known as Cerro Rico de Potosí. This mountain is partially composed of precious metal-polymetallic tin ores. Mining of Potosí ores began in 1545 and has led to the severe contamination of the Pilcomayo River water and sediments. Although toxic waste spills are released daily in the upper basin of the Pilcomayo River, in 1996 and 2005 mine tailings dams collapsed and thousands of tons of toxic wastes have been spread downstream. These toxic spills, which contain high concentrations of arsenic and heavy metals, may severely affect plants, animals, and human health, even several kilometers downstream. For example, in Spain, the Aznalcóllar accident (1998) has severely contaminated the Guadiamar River [3] and the accident at the El Porco mine in Bolivia (in 1996, 50 km from the city of Potosí) contaminated the Pilaya River and part of the Pilcomayo River [4].

Copper is one of the most abundant heavy metals present in the Pilcomayo River water and sediments [5]. Copper is a trace element which is essential to the function of specific proteins and enzymes. However, at high concentrations, it may be toxic to organisms. The toxicity of copper to fish has been well documented. In addition to its acute lethality, a wide range of toxicological responses of several organs to this metal has been reported for various fish species. Copper alters the regular functioning of the gills and liver [6, 7] by causing severe histological changes in these organs. The most frequent physiological effect observed in fish exposed to aqueous copper is ionoregulatory failure [6]. Additionally, aqueous copper has also been reported to influence fish respiration [8, 9]. Biota differences in respiratory physiology, including differences in ventilation rates and volumes, can lead to different internal exposure doses and thus different toxic responses [10]. These differences in respiratory responses to a pollutant might be important.

The impact of copper on the aquatic énvironment is complex and depends on the physicochemical characteristics of water. Alkalinity, hardness, dissolved organic matter, and pH strongly influence copper speciation in water and, consequently, its bioavailability for absorption by fish [11]. Ionic copper (Cu^{2+}) and copper hydroxides are considered the most toxic species of aqueous copper, while copper carbonates have proven much less toxicity [12]. Cu^{2+} is the dominant copper species at pH levels below 7.0, and according to the mean lethal concentration (LC50) ranges in [13] copper is the second most toxic metal to freshwater fish. In soft water, copper is acutely toxic to freshwater teleosts at concentrations between 10 and 20 $\mu g\,L^{-1}$ [14–16] including such cultured species as salmonids, cyprinids, and catfish [17]. Ion-poor and soft waters [18] have a low buffering capacity, and fish culture practices may be accompanied by changes in pH and, hence, in the chemical speciation of copper, potentially increasing its toxic effect. Furthermore, high temperatures tend to increase the diffusion rate, accelerating chemical reactions [19], thereby favoring the toxic action of copper or other heavy metals.

The Pilcomayo River water is characterized by its high concentrations of calcium, sodium, (bi)carbonate, sulphates, and total suspended solids. Water hardness is one of the main and most well recognized of the modifying factors of metal ionic species. Hardness reduces toxicity by "protecting" the organism against metal toxicity via several possible mechanisms [20, 21]. The ameliorative effect of hardness is shown to be more complex than the simple hardness-toxicity relationships would suggest. The hardness cations, calcium and/or magnesium, and protons are thought to inhibit Cu binding/uptake at the cell surface, via different mechanisms [22]. Alkalinity, on the other hand, affects metal ionic species in water solution through their complexation with carbonates [23, 24]. Additionally, dissolved organic matter binds metal species as well [25].

The conceptual Biotic Ligand Model (BLM) [26] may be considered in terms of three separate components: water chemistry, the binding of the toxic metal species to the biotic ligand, and the relationship between metal binding and the toxic response of the aquatic organism [27]. The BLM has been proposed as a tool to evaluate quantitatively the manner in which water chemistry affects the speciation and biological availability of metals in aquatic systems [27]. The toxicology of metals would not be complete without an evaluation of which chemical species are the most toxic and how toxicity might be modified by various environmental factors. These mechanisms need to be uncoupled, if their effects are to be incorporated into models such as the BLM [22], which uses physicochemical variables to predict the acute toxicity of metals, such as copper, to freshwater biota on a site-specific basis. At present, BLM, version 2.2.3, has been developed for two species of fish: fathead minnows (Pimephales promelas) and rainbow trout (Oncorhynchus mykiss), for three species of invertebrates:Daphnia magna, Daphnia pulex, and Ceriodaphnia dubia and four metals: copper, cadmium, silver, and zinc. BLM-predicted Cu LC50 values have agreed to estimated LC50 values over a wide range of water quality characteristics [27]. It is implicitly assumed that BLMs can be extrapolated within taxonomically similar groups; that is, BLMs developed for P. promelas can be applied to toxicity data for other fish species, and BLMs for D. magna and C. dubia can be applied to toxicity data for other invertebrates [28]. The basis for a cross-species extrapolation is the assumption that the parameters which describe interactions between cations (notably calcium, magnesium, and protons), the toxic free metal ion (e.g., Cu^{2+}), and the biotic ligands are similar across organisms and that only intrinsic sensitivity varies among species [28].

Daphnia magna acute toxicity tests have been performed in the Pilcomayo River water from Mision La Paz, Argentina, by Natale et al. [29]. But there are no previous toxicity tests on a vertebrate in the Pilcomayo River water. Cnesterodon decemmaculatus (Pisces: Poeciliidae; Jenyns, 1842) is an endemic member of the fish family Poeciliidae with extensive distribution in Neotropical America. The species attains high densities in a large variety of water bodies within the entire La Plata River and other South American basins. Cnesterodon decemmaculatus is a small, viviparous, microomnivorous, benthic-pelagic, nonmigratory fish (maxi-minimum size, ≈25 and 45 mm for ♂♂ and ♀♀, resp.). This species is easy to handle and breed under laboratory conditions. Also, C. decemmaculatus proved to be adequate as test organism, due to its small size, fast growth, and short reproduction period [30]. Furthermore, several reports found this species to be suitable as a test organism in acute and chronic toxicity bioassays. The ranges of tolerance of C. decemmaculatus to many environmental parameters, for example, temperature, salinity, and pH, match the conditions for most toxicity tests. Cnesterodon decemmaculatus is usually found in anoxic or very scarcely oxygenated water bodies as well. Thereby, C. decemmaculatus has been used by several authors in bioassays [31–35]. Pimephales promelas(Pisces: Cyprinidae; Rafinesque, 1820), one of the fish species for which BLM has been developed, is a temperate, holarctic fresh water fish. As well as C. decemmaculatus, it is quite tolerant to turbid, low-oxygenated water bodies and can be found in muddy ponds and streams that might, otherwise, be inhospitable to other species of fish. It can also be found in small rivers. Because of its relative resilience and large number of offspring produced, US EPA guidelines (United States Environmental Protection Agency) outline its use for the evaluation of acute and chronic toxicity of water samples or chemical species in vertebrate aquatic animals [36, 37].

The aims of this study were to (a) assess Cu toxicity (96 h LC50) to C. decemmaculatus in a surface water with high hardness, sodium, sulfate, and chloride concentrations (Pilcomayo River water), (b) apply BLM, version 2.2.3, to predict acute copper toxicity to P. promelas (Cu LC50) under Pilcomayo River water characteristics, (c) compare the predicted Cu LC50 value for P. promelas to the calculated for C. decemmaculatus in the Pilcomayo River water, and, finally, (d) given that Pilcomayo River hydrochemistry is strongly influenced by the hydrological cycle [34], we also analyze the interseasonal variation of the main water quality parameters that influence copper bioavailability and toxicity.

MATERIALS AND METHODS

Study Area

The Pilcomayo River in South America is a tributary to the large La Plata system. Its headwaters are located in Bolivia along the eastern flank of the Central Andes at an elevation of approximately 5,200 m (Figure 1). The river flows in a southeasterly direction for about 670 km until reaching the Chaco Plains along Bolivia's southern border with Argentina. Its total length is 2,426 km, and its basin covers an area of approximately 288,360 km² (Comisión Trinacional del Río Pilcomayo).

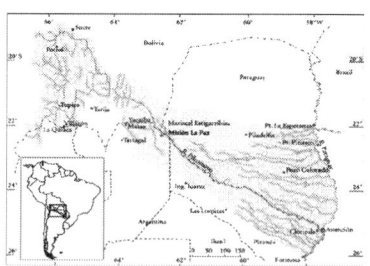

Figure 1: Map of the Pilcomayo River basin with the water sampling location (Misión La Paz, Argentina).

An important feature of the Pilcomayo River, present in all dryland rivers, is its extreme interannual and interseasonal variability in discharge [38]. Interseasonal climatic variation is also extreme in dryland river basins as, frequently, a clearly marked dry and rainy season can be distinguished. These different hydrological regimes are usually associated with important variations in water chemistry and may have important effects on the behavior of aquatic ecosystems and trace metals toxicity.

Water Sampling and Chemical Analysis

Discrete water samples for chemical analyses were taken 10 cm below the water surface and in triplicate from the navigation channel, left and right shore of the Pilcomayo River in the Misión La Paz International bridge (22°22'45"S–62°31'08"W; 254 meters over sea level) in May 2009 (Figure 1). Water sampling took place during the routine water quality monitoring program coordinated by the Subsecretaría de Recursos Hídricos (SsRH-Argentina) and the Comisión Trinacional del Río Pilcomayo. Sampling and in situ water quality determinations were in charge of the SsRH, Centro de Ecología Aplicada del Litoral (CECOAL-CONICET) and Universidad Nacional de Salta (UNS). Laboratory analysis of chemical parameters was performed by the UNS and the Comisión Nacional de Energía Atómica (CNEA-Argentina). Water discharge (Q) was measured by EVARSA-Argentina, pH, and water temperature (T) were determined in situ. Dissolved concentrations of calcium (Ca), magnesium (Mg), chloride (Cl), potassium (K), sodium (Na), sulphate (SO_4), alkalinity (Alk), dissolved organic carbon (DOC), total suspended solids (TSS), total dissolved solids (TDS), and total (T·Cu) and dissolved copper (D·Cu) concentrations were determined using Standard Methods test protocols [39]. Particulate copper (P·Cu) was derived according to the following

$$P \cdot Cu = \frac{[T \cdot Cu] - [D \cdot Cu]}{[TSS]}.$$

(1)

Toxicity Test

Water for the toxicity test was collected in prerinsed 10 L polypropylene containers. Samples were immediately placed into coolers and transported to the laboratory. Later, water was centrifuged (2,000 rpm during 15 minutes) and filtered through 47 mm 0.45 μm pore glass-fiber filters (Whatman GF/C). Copper background concentration in the Pilcomayo river water was 0.02 mg Cu L^{-1}.

Juvenile C. decemmaculatus were collected from a small pond, located in Reserva Natural Los Robles, Buenos Aires Province, Argentina (main chemical and physical parameters are shown in Table 2). Fish were kept at temperatures ranging from 20 to 24°C and pH ranging from pH 7.1 to 7.5 in an aquarium supplied with a continuous flow of aerated de-chlorinated tap water for 30 days. During this period and posterior laboratory and test water (centrifuged and filtered Pilcomayo River water) acclimation, the fish were fed with a daily ration of commercial fish food Shulet. Acclimation to test water (pH of 7.9–8.20, 15–20°C) was performed by adding small quantities of test water to the aquarium until most of the water volume corresponded to test water. One day before and during the experiment, fish were not fed.

Toxicity effect of copper on fish was tested in static systems (4 L glass aquaria) with continuous artificial aeration, constant environmental temperature (20°C), and natural laboratory photoperiod. Test water volume in each aquarium was 2 L. The experimental design included seven different copper concentrations with one control group (kept in test water and without copper addition). Test copper concentrations were attained by spiking from a stock solution of 100 mg Cu L^{-1}. The toxicant used was reagent-grade CuSO$_4$. Dissolved copper concentrations tested were 0.05, 0.19, 0.39, 0.61, 0.73, 1.01, and 1.42 mg Cu L^{-1}.

To define the range of copper concentrations to be employed in the bioassay, a nominal concentration of 0.8 mg Cu L^{-1} was tested in an aquarium with 2 L volume of the Pilcomayo River water and 12 acclimated specimens of juvenile C. decemmaculatus for 96 h. Fish (not sexed) taken from the acclimation tank were randomly distributed in the different experimental aquaria. Mean standard length of the specimens selected was 18.9 mm, and each aquarium contained 10 specimens. Copper concentration in the experimental aquaria was adjusted prior to the fish transfer. Survival was registered four times a day during 96 h. Water pH, conductivity, and dissolved oxygen were measured with portable probes from HANNA (HANNA instruments, Inc. Woonsocket, RI, USA) daily. Water samples were collected into polypropylene conical tubes and acidified to pH<2 with concentrated nitric acid (reagent grade) for metal analysis by atomic absorption spectrophotometry (Perkin Elmer 1100B, Perkin Elmer, Inc., Waltham, MA, USA) after acid digestion (HNO_3:$HClO_4$:HF:HCl). Method copper detection limit was 0.01 mg L^{-1}.

LC50 Calculations

The median lethal concentrations (LC50) at 24, 48, 72, and 96 h (24 h LC50, 48 h LC50, 72 h LC50, 96 h LC50) were calculated using the PROBIT method [40] and the statistical program Statgraphics Plus 5.1 (StatPoint Technologies, Inc., Warrenton, VI, USA).

Version 2.2.3 of the BLM Windows Interface (available at http://www.hydroqual.com/wr_blm.html) was run in order to predict acute copper toxicity to P. promelas (toxicity mode) and copper ionic speciation (speciation mode) on the measured copper concentrations tested. The Pilcomayo River water quality parameters employed to run the BLM were temperature, pH, dissolved organic carbon, calcium, magnesium, sodium, potassium, sulphates, chlorides, alkalinity, and dissolved copper concentrations.

Interseasonal Water Quality Analysis

To determine water discharge influence on major anions and cations and on total and dissolved solids and copper concentrations, we used water quality data, available from 2003 to 2010, from the Misión La Paz monitoring station (provided by La Comisión Trinacional del Río Pilcomayo). Though a large number of water quality parameters are determined, we selected only those that constitute BLM inputs, water discharge, total suspended and dissolved solids, and total and dissolved copper concentrations. Hydrological and water quality data were classified into two groups: data corresponding to the dry season (May–October) and data corresponding to the wet season (November–

April). Distance metric test statistic (*dm*) [41, 42] was calculated in order to determine significant difference between the means. This statistic is defined as the difference between the variables means x and y of the standardized series. Due to the limited data availability for each season, the standard deviations of the errors, ES_x and ES_y, were estimated using bootstrap techniques [43]. Bootstrapping is the practice of estimating properties of an estimator (standard deviations of the errors, in this case) by measuring those properties when sampling from an approximate distribution. It can be implemented by constructing a number of re-samples of the observed dataset (and of equal size to the observed dataset). Each re-sample is obtained by random sampling with replacement from the original dataset.

The *dm* statistic is defined as follows:

$$dm = \frac{\overline{x} - \overline{y}}{\sqrt{ES_x{}^2 + ES_y{}^2}},$$

(2)

where values of |*dm*| higher than 2 are an indication that the corresponding variables means are different.

Pilcomayo River water discharge available data from 1961 to 2008 (provided by SsRH-Argentina) was used to perform the Pilcomayo River hydrograph.

RESULTS

Toxicity Test

No mortality was observed in the control group. An exponential decrease of fish survival with time towards an asymptotic value reached at about 96 h was observed. Figure 2 shows LC50 values as a function of copper exposure time. Data fitted an exponential regression leading to the following equation: LC50 = $1.1428e^{-0.0061t}$ and a R^2 value of 0.9219.

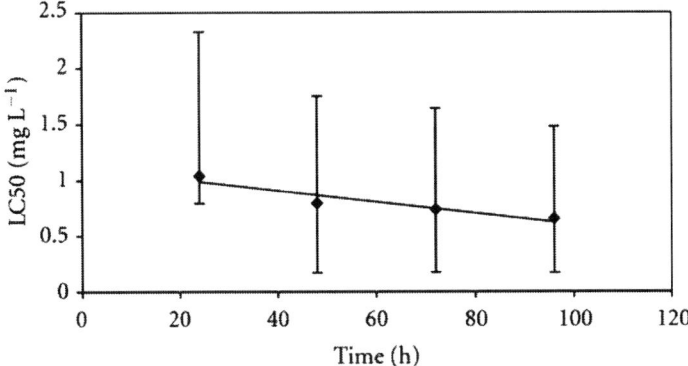

Figure 2: Cnesterodon decemmaculatus copper toxicity test: Cu LC50 values (mg L^{-1}) calculated by PROBIT analysis and confidence intervals (vertical bars) as a function of exposure time (h).

The median lethal concentrations (LC50, mg L^{-1}) at 24, 48, 72, and 96 h (24 h LC50, 48 h LC50, 72 h LC50, 96 h LC50) with their corresponding confidence intervals (quoted) calculated using PROBIT method were 1.039 (1.288–0.245), 0.792 (0.962–0.622), 0.734 (0.908–0.561), and 0.655 (0.823–0.488), respectively (Figure 2).

Biotic Ligand Model

All physicochemical parameters values of the Pilcomayo River water, measured on our sampling date (Table2), were within the range to which BLM can be applied. Calculated 96 h Cu LC50 for C. decemmaculatus was 0.655 mg L^{-1} (0.823–0.488). Predicted 96 h Cu LC50 by BLM developed for P. promelas was 0.722 mg L^{-1}. BLM was also run with water quality data of the test water used by Villar et al. [44] in order to obtain a predicted acute copper toxicity concentration in a soft water. Figure 3 shows that predicted Cu LC50 (µg L^{-1}) was accurate within a factor of 2 for both, hard and soft water (the Pilcomayo River water quality data and Villar et al. [44]).

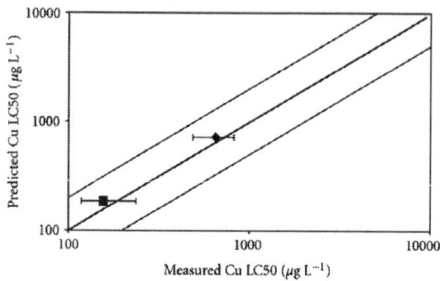

Figure 3: Measured copper toxicity (LC50, in µg L⁻¹) to C. decemmaculatus by Villar et al. (closed square) and the present study (closed diamond) compared with predicted copper toxicity using the BLM developed for P. promelas. The thicker line represents a 1 : 1 relationship. The thinner line represents predictions within a factor of 2. The error bars represent 95% confidence intervals.

BLM copper speciation (Figure 4) shows that $CuCO_3$ is the second most abundant copper chemical species in the control group, in all the concentrations tested up to 0.732 mg L⁻¹ and becomes the most abundant in the last two concentrations, after copper bound to dissolved organic carbon. $CuHCO_3^+$ contribution, amongst the remaining species, is the highest and reaches 27% for the highest copper concentration tested. Moreover, for these two cases, the carbonate fraction of dissolved copper exceeds the organic fraction.

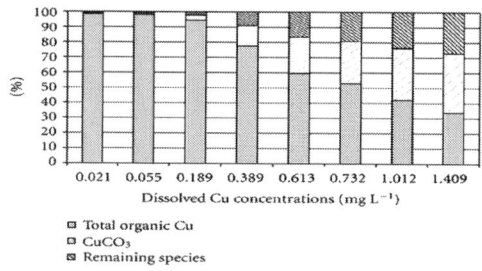

Figure 4: BLM speciation output for each of the copper concentrations tested and control group (first column). Copper species are expressed as percentages of total dissolved copper concentration. Remaining species summarizes the contributions of CuOH, $Cu(OH)_2$, $CuSO_4$, $Cu(CO3)2-2$, CuCl⁺, and CuHCO3+.

Interseasonal Water Quality Analysis

Means, medians, standard deviations, maximum, and minimum values of the selected water quality parameters are shown in Table 1. Water quality of the Pilcomayo River water sampled to perform the bioassay and BLM modeling corresponded to the dry season. For this water sample, pH, TDS, Ca, Mg, Na, K, SO_4, Cl, and Alk were lower than the median values from historical records of the Pilcomayo River in the dry season. However, the respective comparison for water discharge and total suspended solids concentration showed a reverse outcome.

Table 1: Main chemical and physical parameters of the water where C. decemmaculatusspecimens were captured (ND: not detected)

Parameter	
T (°C)	17
pH	7.86
CE (µS)	512
Diss. O2 (mg L−1)	11.02
NH4(mg L−1)	0.005
NO3(mg L−1)	0.006
NO2(mg L−1)	0.003
SRP (mg L−1)	0.141
SO4(mg L−1)	14.2
Cl (mg L−1)	17.5
Alkalinity (mg CaCO3 L−1)	292.9
Mg (mg L−1)	17.8
Ca (mg L−1)	27.7
Cu (mg L−1)	ND
Zn (mg L−1)	0.03
Cr (mg L−1)	0.04
Cd (mg L−1)	ND
Pb (mg L−1)	0.15

Table 2: Seasonal descriptive statistics and *dm* statistic values (difference between season's means) of hydrological and water quality parameters of the Pilcomayo River at Misión La Paz, Argentina (data provided by the Comisión Trinacional del Río Pilcomayo)

	Q m³ s⁻¹	pH UpH	T °C	TSS mgL⁻¹	TDS mgl⁻¹	Ca mgl⁻¹	Mg mgl⁻¹	SO₄ mgl⁻¹	Alk Mg CaCO₃ l⁻¹	Cl mgl⁻¹	Na mgl⁻¹	K mg l⁻¹	DOC mgl⁻¹	T-Cu mgl⁻¹	P-Cu µg kg⁻¹	D-Cu mgl⁻¹
Sampling date (May 2009)	75.20	7.67	21.43	1637	517	73.33	30.5	207.3	110	101	65.2	5.1	4.4	0.07	0.07	0.001
Dry season																
n	14	14	11	13	11	14	14	12	12	12	14	14	6	13	12	14
Mean	57.18	7.87	17.54	1382.95	845.7	80.6	35.22	292.09	126.43	164.91	125.97	6.6	5.27	0.05	61.30	0.0027
Median	53.22	7.91	16.9	1147	847	79	37.75	278	121.5	187.5	120.5	6.55	5.06	0.03	34.37	0.002
Max	161.2	8.1	25.8	8181	1114	150.2	52	399.7	210	243	207	12	10.8	0.18	233.75	0.0067
Min	6.45	7.31	11.8	92	517	44	9.4	190	91	45	70	3.2	2.2	0.01	11.43	0.001
Std	41.40	0.20	4.36	2178.71	211.85	28.64	11.05	71.02	30.39	61.53	43.06	2.29	3.09	0.04	72.5	0.001
Wet season																
n	13	13	8	10	13	13	10	11	11	11	13	13		12	9	13
Mean	248.4	7.70	25.35	14760.2	469.46	51.67	17.35	149.09	108.71	56.85	43.88	5.69		0.16	26.58	0.0025
Median	219.07	7.65	27.15	8990.5	436	44	14.5	135	100	32	36	5		0.11	24.93	0.0012
Max	712.77	8.33	28.2	53960	1168	91	36	376	182	268	131	9.9		0.458	53.78	0.01
Min	6.65	7.27	17.7	240	170	19	3.4	59	57.6	20	18	3.2		0.026	2.724	0.0009
Std	195.22	0.27	3.68	17298.53	255.60	23	10.01	87.87	33.07	71.45	31.33	2.38		0.15	18.75	0.002
dm	-3.8	1.88	-4.36	-2.49	4.07	3.09	4.38	4.53	1.41	4.04	5.79	1.07		-2.70	1.73	0.14

According to dm values (Table 2), it can be seen that water discharge, temperature, calcium, magnesium, sulphates, chlorides, sodium, total suspended and dissolved solids, and total copper concentrations showed interseasonal variation. Although all dm values for these variables are higher than 2, the value itself shows how different the corresponding means are. Total suspended solids and total copper concentrations show interseasonal variation, but the difference between means (dm) is lower compared to other water quality variables. Alkalinity, pH, potassium, particulate copper, and dissolved copper concentrations do not show interseasonal variation (dm values lower than 2). According to dm values, water discharge, temperature, total suspended solids, and total copper are higher during the wet season. The remaining water quality parameters show higher concentrations during the dry season.

The Pilcomayo River hydrograph (Figure 5) is typical of dryland rivers. Water discharge begins to increase on November, peaks on February, and declines gradually reaching the lowest values on September. Mean annual water discharge determined with water discharge record of the last 47 years at Misión La Paz was 212.1 m^3 s^{-1} with a maximum value of 508.7 m^3 s^{-1} and a minimum of 77.7 m^3 s^{-1}. The Pilcomayo River maximum discharge record was registered on March 1984 when water discharge tripled its mean value reaching 1908 m^3 s^{-1}. The corresponding minimum water discharge value of 7.5 m^3 s^{-1} was registered on September 1966. The mean Pilcomayo River water discharge on May 2009, our sampling date (123.6 m^3 s^{-1}), was higher than the historical mean water discharge for May (98.8 m^3 s^{-1}, data not shown) and corresponded to the 75th percentile.

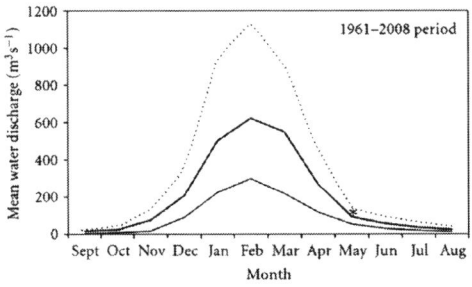

Figure 5: The Pilcomayo River hydrograph. The thicker highlighted line represents mean river water discharge. The dotted line represents the 90th per-

centile water discharge values and the lower thinner line the 10th percentile water discharge values. The asterisk indicates mean water discharge on May 2009 (water sampling), $123.6\,m^3\,s^{-1}$. (Data from Misión La Paz monitoring station, Argentina.).

DISCUSSION

The Pilcomayo River water is very hard surface water. Water hardness is mainly produced by calcium and magnesium concentrations [39]. Copper toxicity to several aquatic species has been reported to be negatively correlated with hardness, but other reports have indicated little or no effect [23]. The effect of hardness on copper toxicity might reflect competition between hardness ions and copper for binding sites on gill surface. Calcium appears to be more protective than magnesium against copper toxicity to fish [45]. Calcium binds to the gill surface and controls the permeability of the membrane and the integrity of the ionoregulatory function [46]. There is only one previous measure of copper toxicity to C. decemmaculatus. Villar et al. [44] found for adults of C. decemmaculatus a 96 h Cu LC50 of $0.155\,mg\,L^{-1}$ in a synthetic soft water with a hardness of $67.66\,mg\,CaCO_3\,L^{-1}$. Acute copper toxicity estimates from this study and Villar et al. [44] were normalized to a hardness of $50\,mg\,CaCO_3\,L^{-1}$ using the US EPA conversion formula for normalization of data given in the ambient water quality criteria for copper (LC50 at 50 mg/L = $e^{\ln(LC50)-0.9422\times(\ln(hardness)-\ln(50))}$) [47]. Normalization of this study toxicity estimates gave a LC50 of $0.12\,mg\,L^{-1}$ (0.08–0.15), and for Villar et al. [44] the normalized LC50 was $0.12\,mg\,L^{-1}$ (0.09–0.18). Although test water used by Villar et al. [44] had lower dissolved organic carbon and alkalinity concentrations, hardness seems to have a strong influence on copper toxicity to C. decemmaculatus. Van Genderen et al. [45] found that increments in water hardness from 200 to $1000\,mg\,CaCO_3\,L^{-1}$, achieved by increasing concentration of calcium (magnesium held constant at $30\,mg\,L^{-1}$), increased 96 h LC50 to larval P. promelas. Some studies have suggested that the molar ratio between calcium and magnesium may be more important than their absolute concentrations. The calcium-to-magnesium molar ratio in the Pilcomayo River water is 1.45, and studies reported that hardness consisting primarily of calcium (molar ratios of >1) is protective of both fish [19, 23] and invertebrates [45]. However, hardness consisting primarily of magnesium (Ca : Mg molar ratios of ≤1) has only been shown to be important for invertebrates [22].

Alkalinity affects copper speciation in solution through complexation with carbonates, which will influence bioavailability [22]. The effects of hardness on aquatic biota toxicity due to metals in some cases are misinterpreted by correlations with alkalinity, pH, and/or other ionic constituents. Van Genderen et al. [45] found that the relationship between alkalinity and LC50 values for P. promelas in the natural waters tested was not significant, but analysis of laboratory water quality data demonstrated a significant positive correlation. Lauren and MacDonald [24] similarly concluded that cupric ion and copper hydroxo complexes, but not copper carbonate complexes, were toxic. When alkalinity is increased, while maintaining a constant pH, copper toxicity has been reported to decrease, but the magnitude of this effect varies with hardness and other experimental conditions and is sometimes not observed [23]. These results show the strong influence of alkalinity on copper bioavailability as copper concentration increases.

In the present study, pH varied from 7.90 to 8.30 in all treatments. Erickson et al. [23] found a decrease in copper toxicity to early-life-stage P. promelas when pH was increased from 6.5 to 8.5–9 in ambient alkalinity ($45\,mg\,CaCO_3\,L^{-1}$) as well as in elevated alkalinity ($150\,mg\,CaCO_3\,L^{-1}$). On the other hand, Lauren and MacDonald [24] concluded that alkalinity, but not pH, affected short-term lethality of copper to rainbow trout. Carvalho and Fernandes [19] found that copper toxicity to Prochilodus scrofa is dependent on water pH. They found lower copper toxicity at pH of 4.5. The stimulation, at low pH, of gill secretion of mucus, which can bind copper, might contribute to the antagonism of low pH with copper toxicity [23]. To perform their toxicity tests, these authors used soft low-alkalinity water and is possible that the reduced concentration of protons and the low levels of calcium in ion-poor soft waters may favor Cu^{2+} binding to the gill surface membrane, increasing the uptake of copper and, hence, its toxicity in high water pH [12].

Erickson et al. [23] found that the addition of potassium chloride increased copper toxicity, while addition of calcium chloride and sodium chloride reduced it, and magnesium chloride had no effect. When calcium, sodium, and magnesium were added as sulfate salts, the same effects were observed. The primary effect of copper is on sodium and chloride uptake and efflux [48]. Exposure to copper produces the inhibition of the active uptake of sodium. In addition, at high enough concentrations, it may also affect the efflux, but this effect would mainly be mediated by a general disruption in gill epithelial integrity. The effects on ionoregulation result in a decrease in levels of plasma sodium, chloride, and other ions, which in turn leads to cardiovascular collapse and death. There is the hypothesis that Cu^{2+} is reduced to Cu^+ by reductases on the cell surface to facilitate uptake via sodium transporters [22]; this might explain the reason why the addition of external sodium and chloride is expected to reduce copper toxicity. It is possible that high sodium and chloride concentrations found in the Pilcomayo River water have some protective effect against copper toxicity. Chloride may also have an effect on copper speciation. However, in BLM copper speciation output (Figure 4), copper chloride is included within the remaining species and its contribution is the lowest. Consequently, in this study, chloride effect on copper bioavailability is negligible.

Other water quality parameters that are considered for their ability to ameliorate copper toxicity by decreasing copper bioavailability are the complexing ligands (dissolved organic carbon, hydroxide, and carbonate) [42]. Matsuo et al. [25] showed that dissolved organic matter forms complexes with Cu^{2+}, which reduces the free form in water and therefore the amount of ionic Cu^{2+} available to bind to the gill sites. These authors concluded that dissolved organic matter has direct effects on the gills because it complexes Cu^{2+} and acts on the transport and permeability properties of the gills. Tao et al. [49] proposed that organic compounds with metals bound may adhere to the mucus of the epithelial cell surface during fish aspiration, and afterwards the dissociation of the complex could then release free copper which, in turn, could be transported into the gill tissue. However, the uptake rate of these compounds would be much slower when compared with that of free ionic copper. In the Pilcomayo River water, the effect of dissolved organic matter (measured as dissolved organic carbon) on copper bioavailability is evident. Organic copper is the most abundant species in all treatments except in the highest copper concentration. Paquin et al. [27] argued that strong ligands, such as dissolved organic matter, at the metal concentrations used in acute toxicity applications could reach saturation and do not exert a controlling influence over metal speciation. Bryan et al. [50] found a higher complexation of copper by dissolved organic matter at low total copper concentrations.

They also found that, in the absence of dissolved organic matter and at pH of 8.5, complexation by carbonate species is considerable, but where the complex $CuCO_3$, rather than $CuHCO_3^+$, is dominant. In our study, BLM shows that there is a reduction in the percentage of copper bound to organic matter and an increment in $CuCO_3$ and secondary in $CuHCO_3^+$, as copper concentration increases.

Few studies have examined the effects of suspended solids on copper toxicity. Erickson et al. [23] results suggested that copper adsorbed onto suspended solids could not be considered to be strictly nontoxic. Tao et al. [51] proposed a mechanism of particulate metal uptake by fish, by desorption of the metal from the particles within the gill microenvironment where the particles adhered to mucus. Natale et al. [29] found higher copper toxicity to D. magna in unfiltered Pilcomayo River water samples compared to toxicity test performed with filtered Pilcomayo River water. The authors attributed the difference to the presence of the suspended solids themselves and/or to bioavailability of toxicants (i.e., copper and other metals) adsorbed onto the particles. To avoid these effects of total suspended solids and for the purpose of comparing the experimental results of the copper toxicity bioassay with the corresponding BLM (which considers that metal bound to particulate matter exerts no toxicity) estimates, test water in our study was centrifuged and filtered. Consequently, the experimental approach employed in this study was not able to provide evidence on the effects of total suspended solids on copper toxicity. If copper bound to total suspended solids is nontoxic to fish species, toxicity measured on the basis of dissolved copper in tests performed with unfiltered Pilcomayo River water should not differ from our results.

In a dryland river basin, as Pilcomayo River, it is expected that differences in water discharge values between the dry and the wet season would influence dissolved concentrations of the water quality variables that determine copper bioavailability and toxicity. During the dry season, higher dissolved calcium, magnesium, sodium, and chloride concentrations may reduce copper toxicity to fish while the opposite is expected during the wet season when dissolved concentrations decrease by effect of dilution. Our water sampling was performed at the onset of the dry season, when dissolved concentrations of major ions begin to increase. Therefore, a higher protective effect of these ions should be observed in a study conducted in the Pilcomayo River water collected at lower water discharges. During the dry season, water hardness can reach values between 400 and 500 mg $CaCO_3$ L^{-1} however, its median value is 332.5 mg $CaCO_3$ L^{-1}. Even though BLM was developed from tests generated in soft and moderately hard waters ($\leq$250 mg $CaCO_3$ L^{-1}), previous studies have suggested that BLM predictions are still accurate in very hard surface waters [45]. During the wet season, water hardness falls to a median value of 184.6 mg $CaCO_3$ L^{-1}.

Temporal variation of dissolved organic carbon was impossible to analyze due to lack of historical data, but our results show a possible saturation of dissolved organic carbon binding sites as copper concentration increases, leading to an increment in carbonate species. Alkalinity did not show temporal variation in its concentration. The effect it exerts on copper bioavailability between seasons will depend more on dissolved organic carbon and dissolved copper concentrations than on alkalinity concentration itself.

The Pilcomayo River high load of suspended solids originates in the erosion of soils in the upper mountainous region of the basin during the rainy season. When toxic waste spill from mine tailings is released into the river, copper adsorbs onto suspended solids and sediment. During the rainy season, sediments and solids resulting from eroding soils are carried downstream. Table 1 shows that the highest percentage of total copper concentration belongs to copper adsorbed onto suspended solids. Thus, during the dry season, lower water discharges promote sedimentation leading to lower total suspended solids and consequently lower total copper concentrations in the water column.

Dissolved copper concentration did not show interseasonal variation. This means that the differences in water discharge values between the dry and the wet seasons would not influence dissolved copper concentrations. Also particulate copper concentration did not show temporal variation. Based on the Lu and Allen approach [52], we calculated the partition coefficient ($Kd=P\cdot Cu/D\cdot Cu$) for each season and the dm statistic. According to the dm value (lower than 2, data not shown), Kd does not show interseasonal variation. The partitioning of copper onto suspended particulate matter of rivers depends on many factors including the solid amount. According to Lu and Allen [52], when total suspended solids are high (100 mg L^{-1}), Kd can be considered to be independent of copper concentration. These authors also found lower Kd values with increasing total suspended solids concentration and that this decrease was less at higher total suspended solids concentration.

BLM-MONTE model [53] estimates K (Kd=1.04×10^6×TSS$^{-0.7436}$) showing this copper-particulates inverse relationship. Total suspended solids values in the Pilcomayo River are much higher, reaching a value of more than 50,000 mg L^{-1}. Therefore, in this extreme case, Kd and dissolved copper concentration could be independent of total suspended solids concentration. We could assume that, given that particulate copper showed no variation between the dry and the wet seasons, the copper associated to binding sites is always the same. Attention should be given to the fact that dissolved copper concentrations records in the Pilcomayo River are quite under fish acute toxicity levels along the entire hydrological cycle.

CONCLUSIONS

We can conclude that both P. promelas and C. decemmaculatus fish species respond similarly to copper, and a cross-species extrapolation of Cu BLM is valid within the Pilcomayo River water quality characteristic parameters and experimental conditions of this toxicity test. For a complete cross-fish-species extrapolation, acute copper toxicity tests across a wide range of water quality conditions should be performed to determine if Cu BLM has the ability to account for differences in toxicity to the fish species tested under various site-specific differences in water quality characteristic parameters.

This study shows the importance of studying temporal variation in water quality variables to derive accurate water quality criteria for toxic metals. In the Pilcomayo River, several water quality parameters related to copper toxicity (calcium, magnesium, sodium, and chlorides) vary significantly from the wet season to the dry season and so the protective effect they can exert on fish. On the other hand, the very high load of suspended solids seems to play an important role in determining copper bioavailability and toxicity, since most of the copper appears adsorbed onto these solids and only a small fraction keeps dissolved and almost invariable between the high and the low water discharge seasons.

ACKNOWLEDGMENTS

This research was supported by grant of the University of Buenos Aires (UBACyT 20020100100135). The authors want to thank EVARSA and Subsecretaría de Recursos Hídricos-Argentina (SsRH), who kindly performed water sampling and monitoring operations at Misión La Paz, Comisión Trinacional del Río Pilcomayo for providing additional Pilcomayo River water quality data and Mrs. Amalia González for the artwork. The authors also want to thank Dr. Sergio Goméz and Dr. Jimena González Naya for useful technical suggestions.

REFERENCES

1.   E. Donadio, "Ecólogos y mega-minería, reflexiones sobre porqué y cómo involucrarse en el conflicto minero-ambiental," Ecología Austral, vol. 19, no. 3, pp. 247–254, 2009.

2.   K. A. Hudson-Edwards, M. G. Macklin, J. R. Miller, and P. J. Lechler, "Sources, distribution and storage of heavy metals in the Río Pilcomayo, Bolivia," Journal of Geochemical Exploration, vol. 72, no. 3, pp. 229–250, 2001.

3.   J. O. Grimalt, M. Ferrer, and E. Macpherson, "The mine tailing accident in Aznalcollar," Science of the Total Environment, vol. 242, no. 1–3, pp. 3–11, 1999.

4.   J. Garcia-Guinea and M. Huascar, "Mining waste poisons river basin," Nature, vol. 387, no. 6629, p. 118, 1997.

5.   A. J. P. Smolders, R. A. C. Lock, G. Van der Velde, R. I. Medina Hoyos, and J. G. M. Roelofs, "Effects of mining activities on heavy metal concentrations in water, sediment, and macroinvertebrates in different reaches of the Pilcomayo River, South America," Archives of Environmental Contamination and Toxicology, vol. 44, no. 3, pp. 314–323, 2003.

6.   C. S. Carvalho and M. N. Fernandes, "Effect of copper on liver key enzymes of anaerobic glucose metabolism from freshwater tropical fish Prochilodus lineatus," Comparative Biochemistry and Physiology A, vol. 151, no. 3, pp. 437–442, 2008.

7.   C. Dautremepuits, S. Paris-Palacios, S. Betoulle, and G. Vernet, "Modulation in hepatic and head kidney parameters

of carp (Cyprinus carpio L.) induced by copper and chitosan," Comparative Biochemistry and Physiology, vol. 137, no. 4, pp. 325–333, 2004.

8. M. W. Beaumont, P. J. Butler, and E. W. Taylor, "Exposure of brown trout Salmo trutta to a sublethal concentration of copper in soft acidic water: effects upon gas exchange and ammonia accumulation,"Journal of Experimental Biology, vol. 206, no. 1, pp. 153–162, 2003.

9. H. A. Campbell, R. D. Handy, and D. W. Sims, "Increased metabolic cost of swimming and consequent alterations to circadian activity in rainbow trout (Oncorhynchus mykiss) exposed to dietary copper,"Canadian Journal of Fisheries and Aquatic Sciences, vol. 59, no. 5, pp. 768–777, 2002.

10. G. De Boeck, W. Meeus, W. D. Coen, and R. Blust, "Tissue-specific Cu bioaccumulation patterns and differences in sensitivity to waterborne Cu in three freshwater fish: rainbow trout (Oncorhynchus mykiss), common carp (Cyprinus carpio), and gibel carp (Carassius auratus gibelio)," Aquatic Toxicology, vol. 70, no. 3, pp. 179–188, 2004.

11. R. C. Playle, R. W. Gensemer, and D. G. Dixon, "Copper accumulation on gills of fathead minnows: influence of water hardness, complexation and pH of the gill micro-environment," Environmental Toxicology and Chemistry, vol. 11, no. 3, pp. 381–391, 1992.

12. J. P. Meador, "The interaction of pH, dissolved organic carbon, and total copper in the determination of ionic copper and toxicity," Aquatic Toxicology, vol. 19, no. 1, pp. 13–32, 1991.

13. C. J. Kennedy, "The toxicology of metals in fishes," in Encyclopedia of Fish Physiology: From Genome to Environment, A. P. Farrell, Ed., vol. 3, pp. 2061–2068, Academic Press, San Diego, Calif, USA, 2011.

14. J. C. A. Marr, J. Lipton, D. Cacela, J. A. Hansen, J. S. Meyer, and H. L. Bergman, "Bioavailability and acute toxicity of copper to rainbow trout (Oncorhynchus mykiss) in the presence of organic acids simulating natural dissolved organic carbon," Canadian Journal of Fisheries and Aquatic Sciences, vol. 56, no. 8, pp. 1471–1483, 1999.

15. J. C. McGeer, C. Szebedinszky, D. G. McDonald, and C. M. Wood, "The role of dissolved organic carbon in moderating the bioavailability and toxicity of Cu to rainbow trout during chronic waterborne exposure," Comparative Biochemistry and Physiology, vol. 133, no. 1-2, pp. 147–160, 2002.P. G. Welsh, J. L. Parrott, D. G. Dixon, P. V. Hodson, D. J. Spry, and G. Mierle, "Estimating acute copper toxicity to larval fathead minnow (Pimephales promelas) in soft water from measurements of dissolved organic carbon, calcium, and pH," Canadian Journal of Fisheries and Aquatic Sciences, vol. 53, no. 6, pp. 1263–1271, 1996.

16. W. A. Wurts and P. W. Pershbacher, "Effects of bicarbonate alkalinity and calcium on the acute toxicity of copper to juvenile channel catfish (Ictalurus punctatus)," Aquaculture, vol. 125, no. 1-2, pp. 73–79, 1994.

17. J. Schjolden, J. Sørensen, G. Nilsson, and A. Poléo, "The toxicity of copper to crucian carp (Carassius carassius) in soft water," Science of the Total Environment, vol. 384, no. 1–3, pp. 239–251, 2007.

18. C. S. Carvalho and M. N. Fernandes, "Effect of temperature on copper toxicity and hematological responses in the neotropical fish Prochilodus scrofa at low and high pH," Aquaculture, vol. 251, no. 1, pp. 109–117, 2006.

19. J. B. Sprague, "Factors that modify toxicity," in Fundamentals of Aquatic Toxicology, G. M. Rand and S. R. Petrocelli, Eds., pp. 124–163, Hemisphere Publishing, Washington, DC, USA, 1985.

20. C. M. Wood, W. J. Adams, and G. T. Ankley, "Environmental toxicology of metals," in Reassessment of Metals Criteria for Aquatic Life Protection: Priorities for Research and Implementation, H. L. Bergman and E. J. Dorward-King, Eds., pp. 31–55, SETAC Press, Pensacola, Fla,USA, 1997.

21. S. J. Markich, A. R. King, and S. P. Wilson, "Non-effect of water hardness on the accumulation and toxicity of copper in a freshwater macrophyte (Ceratophyllum demersum): how useful are hardness-modified copper guidelines for protecting freshwater biota?" Chemosphere, vol. 65, no. 10, pp. 1791–1800, 2006.

22. R. J. Erickson, D. A. Benoit, V. R. Mattson, H. P. Nelson, and E. N. Leonard, "The effects of water chemistry on the toxicity of copper

to fathead minnows," Environmental Toxicology and Chemistry, vol. 15, no. 2, pp. 181–193, 1996.

23. D. J. Lauren and D. G. McDonald, "Influence of water hardness, pH, and alkalinity on the mechanisms of copper toxicity in juvenile rainbow trout, Salmo gairdneri," Canadian Journal of Fisheries and Aquatic Science, vol. 43, pp. 1488–1496, 1986.

24. A. Y. O. Matsuo, R. C. Playle, A. L. Val, and C. M. Wood, "Physiological action of dissolved organic matter in rainbow trout in the presence and absence of copper: sodium uptake kinetics and unidirectional flux rates in hard and softwater," Aquatic Toxicology, vol. 70, no. 1, pp. 63–81, 2004.

25. D. M. Di Toro, H. E. Allen, H. L. Bergman, J. S. Meyer, P. R. Paquin, and R. C. Santore, "Biotic ligand model of the acute toxicity of metals. 1. Technical basis," Environmental Toxicology and Chemistry, vol. 20, no. 10, pp. 2383–2396, 2001.

26. P. R. Paquin, J. W. Gorsuch, S. Apte et al., "The biotic ligand model: a historical overview," Comparative Biochemistry and Physiology C, vol. 133, no. 1-2, pp. 3–35, 2002.

27. E. C. Schlekat, E. Van Genderen, K. A. C. De Schamphelaere, P. M. C. Antunes, E. C. Rogevich, and W. A. Stubblefield, "Cross-species extrapolation of chronic nickel Biotic Ligand Models," Science of the Total Environment, vol. 408, no. 24, pp. 6148–6157, 2010.

28. O. E. Natale, C. E. Gómez, and M. V. Leis, "Application of the Biotic Ligand model for regulatory purposes to selected rivers in Argentina with extreme water-quality characteristics," Integrated Environmental Assessment and Management, vol. 3, no. 4, pp. 517–528, 2007.

29. R. C. Menni, Monografías del Museo Argentino de Ciencias Naturales, vol. 5, 2004.

30. F. R. de La Torre, S. O. Demichelis, L. Ferrari, and A. Salibián, "Toxicity of Reconquista river water: bioassays with juvenile Cnesterodon decemmaculatus," Bulletin of Environmental Contamination and Toxicology, vol. 58, no. 4, pp. 558–565, 1997.

31. F. R. de la Torre, L. Ferrari, and A. Salibián, "Freshwater pollution biomarker: response of brain acetylcholinesterase activity in two

fish species," Comparative Biochemistry and Physiology C, vol. 131, no. 3, pp. 271–280, 2002.

32. F. R. de La Torre, L. Ferrari, and A. Salibián, "Biomarkers of a native fish species (Cnesterodon decemmaculatus) application to the water toxicity assessment of a peri-urban polluted river of Argentina," Chemosphere, vol. 59, no. 4, pp. 577–583, 2005.

33. L. Ferrari, M. E. García, F. R. de la Torre, and S. O. Demichelis, "Evaluación Ecotoxicológica del agua de un río urbano mediante bioensayos con especies nativas," Revista del Museo Argentino de Ciencias Naturale, vol. 148, pp. 1–16, 1998.

34. S. Gómez, C. Villar, and C. Bonetto, "Zinc toxicity in the fish Cnesterodon decemmaculatus in the Paraná River and Rio de la Plata Estuary," Environmental Pollution, vol. 99, no. 2, pp. 159–165, 1998.

35. "Pimephales promelas," in FishBase, Froese, Rainer, and D. Pauly, Eds., 2006.

36. J. R. Quinn, Our Native Fishes: The Aquarium Hobbyist's Guide to Observing, Collecting, and Keeping Them, The Countryman Press, Woodstock, VT, USA, 1990.

37. A. J. P. Smolders, K. A. Hudson-Edwards, G. Van der Velde, and J. G. M. Roelofs, "Controls on water chemistry of the Pilcomayo river (Bolivia, South-America)," Applied Geochemistry, vol. 19, no. 11, pp. 1745–1758, 2004.

38. American Public Health Association, American Water Works Association, and Water Environment Federation, Standard Methods for the Examination of Water and Wastewater, American Public Health Association, American Water Works Association, Water Environment Federation, Washington, DC, USA, 20sth edition, 2000.

39. D. J. Finney, Statistical Method in Biological Assay, Charles Griffin, London, Uk, 1978.

40. W. K. Newey and K. D. West, "Hypothesis testing with efficient method of moments estimation,"International Economic Review, vol. 28, pp. 777–787, 1987.

41. W. K. Newey and D. McFadden, "Chapter 36 large sample estimation and hypothesis testing,"Handbook of Econometrics, vol. 4, pp. 2111–2245, 1994.

42. B. Efron and R. J. Tibshirani, An Introduction to the Bootstrap, Chapman & Hall, London, UK, 1993.

43. C. A. Villar, S. E. Goméz, and C. A. Bentos, "Lethal concentration of Cu in the neotropical fishCnesterodon decemmaculatus(Pisces, Cyprinodontiformes)," Bulletin of Environmental Contamination and Toxicology, vol. 65, no. 4, pp. 465–469, 2000.

44. E. Van Genderen, R. Gensemer, C. Smith, R. Santore, and A. Ryan, "Evaluation of the Biotic Ligand Model relative to other site-specific criteria derivation methods for copper in surface waters with elevated hardness," Aquatic Toxicology, vol. 84, no. 2, pp. 279–291, 2007.

45. J. B. Hunn, "Role of calcium in gill function in freshwater fishes," Comparative Biochemistry and Physiology—Part A, vol. 82, no. 3, pp. 543–547, 1985.

46. US Environmental Protection Agency Office of Water Regulations Standards Criteria Standards Division, Ambient Water Quality Criteria for Copper-1984, Washington, DC, USA, 1985.

47. D. J. Lauren and D. G. McDonald, "Effects of copper on branchial ionoregulation in the rainbowtrout,Salmo gairdneri Richardson," Journal of Comparative Physiology, vol. 155, pp. 635–644, 1985.

48. S. Tao, S. Xu, J. Cao, and R. Dawson, "Bioavailability of apparent fulvic acid complexed copper to fish gills," Bulletin of Environmental Contamination and Toxicology, vol. 64, pp. 221–227, 2000.

49. S. E. Bryan, E. Tipping, and J. Hamilton-Taylor, "Comparison of measured and modelled copper binding by natural organic matter in freshwaters," Comparative Biochemistry and Physiology C, vol. 133, no. 1-2, pp. 37–49, 2002.

50. S. Tao, T. Liang, C. F. Liu, and S. P. Xu, "Uptake of copper by neon tetras (Paracheirodon innesi) in the presence and absence of particulate and humic matter," Ecotoxicology, vol. 8, no. 4, pp. 269–275, 1999.

51. Y. Lu and H. E. Allen, "Partitioning of copper onto suspended particulate matter in river waters,"Science of the Total Environment, vol. 277, no. 1–3, pp. 119–132, 2001.

52. Hydroqual, BLM-MONTE User's Guide, Version 2.0, Mahawah, NJ, USA, 2001, 07430.

Chapter 2

Occurrence, Risk Assessment, and Source Apportionment of Heavy Metals in Surface Sediments from Khanpur Lake, Pakistan

Javed Iqbal and Munir H Shah

Department of Chemistry, Quaid-i-Azam University, Islamabad 45320, Pakistan

ABSTRACT

Background

The present study was carried out to assess the seasonal variations, source apportionment, and risk assessment of heavy metals (Cd, Cr, Cu, Fe, Mn, Pb, and Zn) in the surface sediments from the Khanpur Lake, Pakistan.

Methods

Composite samples are collected and processed to measure the concentrations of heavy metals in $Ca(NO_3)_2$ extract and acid extract of the sediments using flame atomic absorption spectrophotometry.

Results

The highest concentrations in acid extracts of the sediments are found for Fe, followed by Mn, while the least concentrations are noted for Cd. Relatively higher extraction efficiencies in $Ca(NO_3)_2$ extract are observed for Pb and Cd, which also reveal extremely severe enrichment in the sediments as shown by the enrichment factor. Geoaccumulation index shows moderate and strong to extreme pollution of Pb and Cd, respectively, whereas potential ecological risk factor exhibits low to very high risk by Cd; the cumulative ecological risk index reveals low to very high risk of contamination in the sediments as a whole. Principal component analysis and cluster analysis reveal dominant anthropogenic contributions of Cd, Pb, Cr, and Zn.

Conclusions

Measured concentrations of Cd, Cr, Cu, Mn, and Pb in the sediments exceed the sediment quality guideline for the lowest effect levels (LEL), while the concentrations of Cd and Pb are also higher than the effects range low (ERL) values, manifesting occasional adverse biological effects to the surrounding flora and fauna. Moreover, the mean effects range medium (ERM) quotient reveals 21% probability of toxicity in the sediments.

BACKGROUND

Contamination of aquatic ecosystems with heavy metals has received much attention due to their toxicity, abundance, and persistence in the environment and subsequent accumulation in aquatic habitats (Arnason and Fletcher [2003]). Elevated levels of heavy metals in environmental compartments, such as aquatic sediments, may pose a risk to human health due to their transfer in aquatic media and uptake by living organisms, thereby entering the food chain (Sin et al. [2001]; Varol and Sen [2012]). Heavy metals may enter a freshwater reservoir from a variety of sources, either natural or anthropogenic (Adaikpoh et al. [2005]; Akoto et al. [2008]). Generally, in natural ecosystems, most of the metals are present in very low concentrations and are mostly derived from rock and soil weathering (Reza and Singh [2010]; Varol and Sen [2012]). Major anthropogenic sources of heavy metal pollution are mining and smelting activities, atmospheric deposition, disposal of untreated/partially treated urban and industrial effluents, metal chelates from different industries, and haphazard use of heavy metal-containing fertilizers and pesticides during agricultural activities (Martin [2000]; Nouri et al. [2008]; Reza and Singh [2010]).

Sediments are ecologically sensitive components of the aquatic ecosystems and are also a reservoir of the contaminants, which take part considerably in maintaining the trophic status for any water reservoir (Singh et al. [2005]). Depending upon the physicochemical conditions, sediments can act both as source and sink for nutrients and heavy metals. Hence, sediments are not only considered as carriers of contaminants but also potential secondary sources of contaminants in an aquatic ecosystem. Consequently, the analysis of sediments is a useful method to study the heavy metal pollution in any area (Gielar et al. [2012]; Varol and Sen [2012]). The toxicity and mobility of the metals in sediments vary among different chemical forms (Cuong and Obbard [2006]; Yu et al.[2010]). Therefore, the evaluation of distribution and mobility/potential bioavailability of heavy metals in surface sediments is an important step to evaluate the degree of contamination of an aquatic ecosystem (Martin et al. [2009]; Sprovieri et al. [2007]). Assessment of biologically available fractions of heavy metals helps to evaluate their potential for mobilization and availability to benthic organisms (Rodrigues et al. [2010]). Various chemical extraction methods have been suggested to determine the bioavailable fractions of the metals in sediments. Generally, weak acids/electrolytes are used to extract the bioavailable fractions of the metals in sediments (An and Kampbell [2003]).

Major objectives of the present study are (i) to measure the concentrations of heavy metals (Cd, Cr, Cu, Fe, Mn, Pb, and Zn) in sediments during summer and winter; (ii) to determine potential ecological risk using enrichment factor (EF), geoaccumulation index

(I_{geo}), potential ecological risk factor (E_i), and potential ecological risk index (RI); (iii) to identify risks of potential toxicity by comparison with sediment quality guidelines (SQGs); (iv) to determine potential bioavailability and mobility of the metals; and (v) to define their natural/ anthropogenic contributions using multivariate statistical methods. It is anticipated that the study would provide a baseline data regarding the distribution and accumulation of heavy metals in the sediments and would help reduce the contamination by identifying the major pollution sources.

METHODS

Study Area

Khanpur Lake (longitude 72°56′E and latitude 33°48′N) is situated on the Haro river near the town of Khanpur, about 40 km northwest of Islamabad, Pakistan (Figure 1). It supplies drinking water to the inhabitants of twin cities of Islamabad and Rawalpindi, Pakistan, and irrigation water to the agricultural areas surrounding the cities. It was built in 1983 with the storage capacity of 140 million m^3 of water. It is 51 m high with an average depth of 15 m. The gross storage capacity of the reservoir is 0.132 km^3 with a total catchment area of 798 km^2. The surface area of the reservoir varies from maximum of 1,806 ha to minimum of 215 ha. In past, the lake was leased for commercial exploitation. The area around the lake has been planted with flowering trees and laid out with gardens, picnic spots, and secluded paths. The lake is used for picnics, fishing, boating, sailing, water skating, and diving. Untreated and/or partially treated urban and industrial effluents, road and agricultural run offs, poultry farms wastes, and contaminants released during the recreational use of motorboats are among the suspected sources of pollution in the lake.

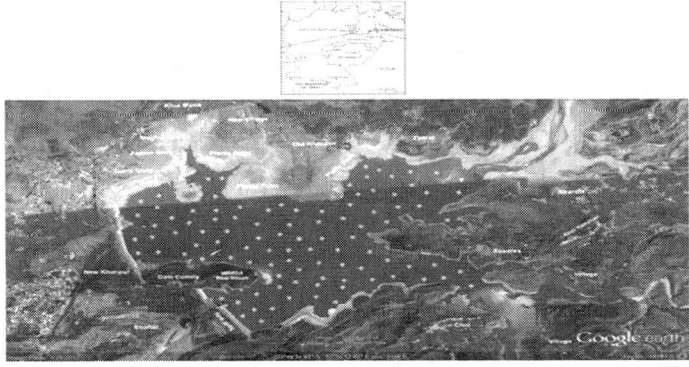

Figure 1: Location of the sampling points in the study area.

Sampling and Preservation

A total of 100 composite surface sediment samples from Khanpur Lake, Pakistan, were collected in the summer and winter of 2008. Each sediment sample was a composite of three to five sub-samples from an area of 1 to 2 m² and collected using a snapper (Ø 5 cm) in top layer (0 to 10 cm). The sediment samples were taken from the central portion of the snapper with a plastic spatula to avoid any contamination from the metallic parts of the sampler. Before transferring the samples in pre-cleaned Ziploc polythene bags (S. C. Johnson & Son, Inc., Racine, WI, USA), the above water was decanted. The samples were kept in airtight large plastic containers for transport to the laboratory. The sediment samples were then oven-dried, grounded, homogenized, and sealed in pre-cleaned polythene bags and stored in a refrigerator until further processing (Radojevic and Bashkin [1999]).

Sample Processing and Analysis

The samples were processed to assess the $Ca(NO_3)_2$-extractable and acid-extractable fractions of heavy metals. A single-step extraction procedure using 0.1 M $Ca(NO_3)_2$ was applied to the sediment samples at room temperature in order to evaluate the bioavailable metal fractions (An and Kampbell[2003]). An aliquot of 5 g of the sample was added to 50-mL solution of 0.1 M $Ca(NO_3)_2$, and the extraction was performed in pre-cleaned glass vessel by shaking on an auto-shaker at 240 vibrations/min for 16 h. A blank sample was also processed with the same amount of reagents without sediment sample. Three replicate extractions were performed for each sample. The final extracts were separated from the solid residues through filtration using a fine (0.45-μm pore) filter paper (An and Kampbell [2003]; Radojevic and Bashkin [1999]; Rodrigues et al. [2010]). To measure the acid-extractable fractions, 1- to 2-g dried sediment sample was digested in a microwave system using an acid mixture of 9 mL HNO_3 and 3 mL HCl (USEPA [2007]). Three replicate extractions were performed for each sample. The digests were then filtered through the fine filter paper and made up to 50 mL with double distilled water and stored at 4°C. A blank sample was also processed with the same amount of chemical reagents without sediment sample. Heavy metals (Cd, Cr, Cu, Fe, Mn, Pb, and Zn) in the sediment samples were analyzed using a flame atomic absorption spectrophotometer (Shimadzu AA-670, Kyoto, Japan). The calibration line method was used for quantification of the metals, and the samples were appropriately diluted whenever required (Radojevic and Bashkin [1999]; Shah et al. [2012]). The optimum analytical conditions used for the quantification of the selected metals on the spectrophotometer are given in Table 1. During sample collection and analysis, strict QA/QC measures were taken including method blanks, analysis of standard reference material, and analysis of duplicate samples. The reagents for the blanks were prepared during each extraction, and all samples were blank-corrected. Standard reference material (NIST SRM-2709) was also used to ensure the reliability of the metal data as shown in Table 1. The measured metal levels closely matched with the certified values. Moreover, reliability of the finished data was also ensured using known spikes and by conducting interlaboratory comparison, and the results were within ±1.5%. Working standards of the metals were prepared from a stock solution of 1,000 mg/L (E-Merck, Darmstadt, Germany) by successive dilutions. The moisture content of each sediment sample was determined by drying separate 5-g sample in an oven (105°C ± 2°C) to constant weight. From this, a correction to dry mass was obtained, which was applied to all reported metal concentrations. All the measurements were made in triplicate.

Table 1: Description of optimum analytical conditions and analysis of selected metals in SRM

	Cd	Cr	Cu	Fe	Mn	Pb	Zn
Wavelength (nm)	228.8	357.9	324.8	248.3	279.5	217.0	213.9
HC lamp current (mA)	4.0	5.0	3.0	8.0	5.0	7.0	4.0
Slit width (nm)	0.3	0.5	0.5	0.2	0.4	0.3	0.5
Fuel gas flow rate (L/min)	1.8	2.6	1.8	2	1.9	1.8	2
Detection limit (µg/L)	4.0	6.0	4.0	6.0	3.0	10.0	2.0
SRM-certified level (mg/kg)	0.38	130	34.6	35,000	538	18.9	106
SRM-measured level ± SD (mg/kg)	0.36 ± 0.03	138 ± 8	35.2 ± 1.2	34,300 ± 385	547 ± 11	19.3 ± 1.4	109 ± 3.2

The analytical conditions were maintained on AAS using air-acetylene flame, and the standard reference material is SRM-2709.

Iqbal and Shah

Iqbal and Shah Journal of Analytical Science and Technology 2014 5:28 doi:10.1186/s40543-014-0028-z

Statistical Analysis

Statistical analysis can be used to evaluate the complex eco-toxicological processes by showing the relationship and interdependency among the variables and their relative weights. Basic statistical parameters, such as minimum, maximum, mean, median, standard error (SE), and skewness, were computed along with correlation study. Multivariate techniques have been used for evaluation and characterization of analytical data (Fadigas et al. [2010]). Principal component analysis (PCA) and cluster analysis (CA) are among the most popular methods. The PCA finds out the diagonalization of the covariance or correlation matrix transforming the original chemical measurements into linear combinations of these measurements, which are the principal components (PCs). It rotates the coordinate space axes so that the explained variance of each PC is maximized. This technique allows for data reduction from higher to lower dimensional spaces to simplify their representation. Nonetheless, CA demonstrates the similarities between variables by examining the interpoint distances representing all possible variables in the higher dimensional space. The PCA was performed using varimax normalized rotation on the dataset, and the CA was applied to the standardized matrix of the samples using Ward's method, and the results are reported in the form of dendrograms. PCA and CA complement each other and have been widely used in environmental studies (Gielar et al.[2012]; Iqbal and Shah [2011]; Shah et al. [2012]; Singh et al. [2005]).

Pollutant Indicators and Risk Assessment

To gauge the degree of contamination and to distinguish natural and anthropogenic inputs, EFs, I_{geo}, Ei, and RI are computed (Cukrov et al. [2011]; Hakanson [1980]; Müller [1969]). EFs are calculated (Cukrov et al. [2011]; Iqbal and Shah [2011]; Luoma and Rainbow [2008]; Tessier et al. [2011]) by comparing the measured metal levels to the pre-industrial levels (Lide [2005]). In order to avoid the overestimation or underestimation of the enrichment; geochemical normalization based on the concentration of a conservative element is commonly employed. The purpose of normalization is to correct changes in the nature of sediments, which may influence the contaminant distribution. Various conservative elements may be used: Al, Fe, Th, Ti, Zr, etc. (Larrose et al. [2010]; Reimann and de Caritat [2005]). Iron is chosen as the conservative element for normalization in this work. The interest of using Fe content is its relationship to the abundance of clay and other aluminum silicates in the sediments. Its contents are influenced by natural sedimentation and the effects of enhanced erosion, but not by pollution (Iqbal and Shah [2011]). The normalized EF is usually computed as double ratios of the target element and Fe as a reference element in the examined sediments and Earth's crust using the following relationship:

$$EF = \frac{[X/Fe]_{sample}}{[X/Fe]_{crust}},$$

(1)

where $[X/Fe]_{sample}$ and $[X/Fe]_{crust}$ refer, respectively, to the ratios of mean concentrations (mg/kg, dry weight) of the target element and Fe in the sediments and continental crust (Lide [2005]).

The I_{geo} enables the assessment of contamination by comparing the measured and pre-industrial concentrations of the metals in the Earth's crust (Loska et al. [2004]; Muller [1969]). It is computed using the following relationship:

$$I_{geo} = \log_2 \left(C_n \Big/ 1.5B_n \right),$$

(2)

where C_n is the measured concentration of the element in the sediment samples, and Bn is the geochemical background value in the Earth's crust (Lide [2005]). Factor 1.5 is introduced to minimize the effect of possible variations in the background values which may be attributed to lithogenic variations.

RI is introduced to assess the degree of heavy metal pollution in sediments, which was originally introduced by Hakanson ([1980]), according to the toxicity of heavy metals and the response of the environment:

$$RI = \sum E_i$$

(3)

$$E_i = T_i f_i$$

(4)

$$f_i = C_i / C_b,$$

(5)

where RI is computed as the sum of all risk factors in sediments, E_i is the monomial potential ecological risk factor for individual factors, and Ti is the metal toxic factor. Based on the standardized heavy metal toxic factor developed by Hakanson ([1980]), the order of the level of heavy metal toxicity is $Cd > Pb = Cu > Cr > Zn$. The toxic factors for the metals are 30, 5, 5, 2, and 1, respectively. fiis the metal pollution factor, C_i is the concentration of metal in the sediments, and Cb is the reference value of a given metal in the Earth's crust (Lide [2005]).

Multiple contamination which is often encountered in natural environments affected by human activities is also calculated in terms of mean-effects range medium-quotient (m-ERM-Q) by the following relationship (de Vallejuelo et al. [2010]; Long and MacDonald [1998]; Tessier et al.[2011]):

$$m-ERM-Q = \frac{\sum_{i=1}^{n} C_i / ERM_i}{n},$$

(6)

where C_i is the concentration of a metal in a sediment, ERM_i is the ERM value for metal i, and n is the number of metals.

RESULTS AND DISCUSSION

Distribution of Heavy Metals in the Sediments

Concentrations of heavy metals in acid extracts of the sediments during summer and winter in terms of statistical distribution parameters are shown in Table 2. During summer, the data reveal dominant mean level of Fe (4,630 mg/kg), followed by Mn (447.5 mg/kg), while the average concentration of Cd (1.883 mg/kg) is the lowest. On the average basis, the metals follow a decreasing concentration order: Fe > Mn > Zn > Cu > Cr > Pb > Cd. Among the metals, Fe indicates almost comparable mean and median levels with lower skewness, indicating relatively symmetrical distribution in acid extract of the sediments. The counterpart statistical data during winter show the highest average levels of Fe (3,791 mg/kg), followed by Mn (321.4 mg/kg), whereas Pb (18.24 mg/kg) and Cd (2.457 mg/kg) are found at relatively lower levels. On the mean basis, the metals exhibit a decreasing concentration order: Fe > Mn > Zn > Cr > Cu > Pb > Cd. Relatively normal distribution is revealed by Cd and Pb, which are also associated with lower skewness. Maximum dispersion in terms of SE is exhibited by Fe. Overall, significantly elevated average levels of the metals (except Cd and Cr) are noticed during summer compared with winter (Table 2). It could be due to the leaching of the metals into the reservoir from the roadside and agricultural runoffs during wet summer season.

Table 2: Statistical summary of heavy metal distribution in acid extract and Ca(NO$_3$)$_2$ extract of the sediments

		Summer (n = 50)					Winter (n = 50)					
		Min	Max	Mean	SE	Skew	Min	Max	Mean	SE	Skew	pvalue
Acid extract	Cd	0.196	4.500	1.883	0.234	0.584	0.149	5.183	2.457	0.235	-0.084	<0.05
	Cr	11.35	63.45	34.66	2.293	-0.262	23.82	68.97	37.65	1.543	1.790	Non-significant
	Cu	25.15	49.39	36.84	1.285	-0.072	18.22	51.53	28.05	1.314	1.166	<0.05
	Fe	3,835	5,186	4,630	57.83	-0.350	3,523	4,182	3,791	30.95	0.426	<0.05
	Mn	236.2	836.7	447.5	32.97	0.713	167.7	886.0	321.4	26.00	2.234	<0.05
	Pb	9.739	78.48	33.71	3.419	0.771	0.412	39.03	18.24	1.966	-0.003	<0.01
	Zn	70.71	114.4	86.09	2.032	0.650	42.24	115.2	61.90	2.459	2.190	<0.05
Ca(NO$_3$)$_2$ extract	Cd	0.004	0.122	0.058	0.006	-0.081	0.016	0.146	0.071	0.006	0.175	Non-significant
	Cr	0.042	0.546	0.217	0.027	0.794	0.008	0.478	0.230	0.023	0.061	Non-significant
	Cu	0.008	0.220	0.098	0.008	0.212	0.012	0.134	0.073	0.006	-0.201	<0.05
	Fe	0.020	28.50	2.069	0.985	4.415	0.248	1.218	0.658	0.053	0.314	<0.01
	Mn	0.004	0.274	0.078	0.012	1.149	0.010	0.072	0.042	0.003	0.178	<0.01
	Pb	0.206	2.032	1.192	0.085	-0.343	0.140	2.082	1.205	0.087	-0.254	Non-significant
	Zn	0.010	0.656	0.179	0.023	1.972	0.056	0.186	0.118	0.007	0.175	<0.05

The heavy metal distribution is expressed in milligrams per kilogram.

Iqbal and Shah

Iqbal and Shah Journal of Analytical Science and Technology 2014 5:28   doi:10.1186/s40543-014-0028-z

Correlation Study

The correlation coefficient matrix of heavy metals in the acid extract of the sediments during summer and winter is given in Table 3. During summer, strong correlations of Fe with Mn and Cu, Cr with Zn, and Cu with Mn are noted. Some other significant relationships of Pb with Cd and Cr are also observed. However, Pb and Zn show negative associations with Cu, Fe, and Mn, revealing their opposing distribution in the sediments during summer. The counterpart data related to the metal levels in the sediments during winter indicate strong correlations of Zn with Cu and Mn, Cu with Mn, and Cr with Cd and Cu, thus manifesting close association of these metals which might share common sources. Some significant correlations for Pb with Cr, Cu, Mn, and Zn are also observed. Fe does not show any significant relationship with other heavy metals in the sediments during winter, suggesting its independent variations in the sediments.

Table 3: Correlation coefficients (r)* matrix for heavy metals in acid extract of sediments during summer and winter

	Cd	Cr	Cu	Fe	Mn	Pb	Zn
Cd	1	0.580	0.389	−0.029	0.473	0.181	0.223
Cr	0.320	1	0.657	0.087	0.319	0.412	0.345
Cu	0.171	0.153	1	0.143	0.685	0.480	0.803
Fe	0.071	−0.056	0.615	1	0.126	0.151	−0.016
Mn	0.086	−0.031	0.863	0.551	1	0.425	0.619
Pb	0.362	0.432	−0.068	−0.187	−0.087	1	0.427
Zn	0.103	0.540	−0.028	−0.208	−0.245	0.062	1

Values for summer are below the diagonal, and those for winter are above the diagonal. *r values >0.330 or <−0.330 are significant at $p < 0.01$.

Iqbal and Shah

Iqbal and Shah Journal of Analytical Science and Technology 2014
5:28   doi:10.1186/s40543-014-0028-z

Pollution Indices

The range and mean EF values of heavy metals in acid extract of the sediments during summer and winter are shown in Figure 2a. Seven degrees of contamination are commonly defined (Birch et al.[2003]): $EF < 1$ indicates no enrichment, $EF < 3$ minor enrichment, $EF = 3$ to 5 moderate enrichment, $EF = 5$ to 10 moderately severe enrichment, $EF = 10$ to 25 severe enrichment, $EF = 25$ to 50 very severe enrichment, and $EF > 50$ extremely severe enrichment. During summer, on the average basis, Cr reveals moderate enrichment, Cu and Mn indicate moderately severe enrichment, Zn manifests severe enrichment, Pb shows very severe enrichment, and Cd illustrates extremely severe enrichment in the sediments. The geochemical normalization study during winter reveals that Cr, Cu, and Mn indicate moderately severe enrichment; Pb and Zn explicate severe enrichment, and Cd illuminates extremely severe enrichment. Overall, Cd emerge as the major pollutant during both seasons; Pb poses severe to extremely severe enrichment during summer and minor to very severe enrichment during winter. Zn causes severe enrichment during both seasons. Mostly, elevated degree of pollution by the metals is noted during summer than during winter.

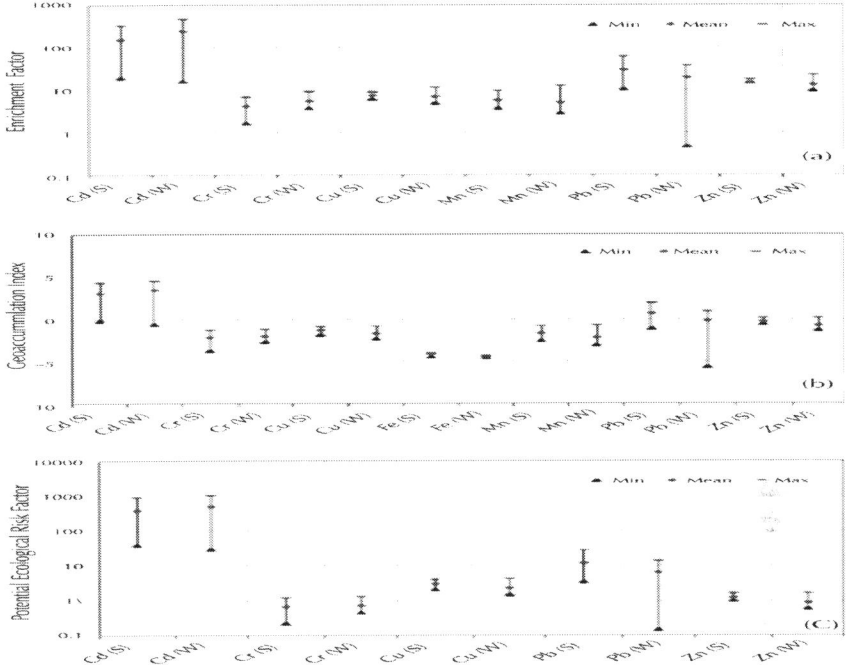

Figure 2: Description of the different parameters. Description of (a)enrichment factor (EF), (b) geoaccumulation index (I_{geo}) and (c) potential ecological risk factor (Ei) of heavy metals in acid extract of sediments during summer (S) and winter (W).

The lowest, mean, and highest values of I_{geo} in acid extract of the sediments during summer and winter are illustrated in Figure 2b. The following categorizations are given by Muller ([1969]) for geoaccumulation index: $I_{geo} < 0$ indicates unpolluted, $I_{geo} = 0$ to 1 unpolluted to moderately polluted, $I_{geo} = 1$ to 2 moderately polluted, $I_{geo} = 2$ to 3 moderately to strongly polluted, $I_{geo} = 3$ to 4 strongly polluted, $I_{geo} = 4$ to 5 strongly to extremely polluted, and $I_{geo} > 5$ demonstrates extremely polluted. The highest category reflects at least a 100-time enrichment above the background values. As shown in the figure, during summer, Cd and Pb pose strong to extreme contamination and moderate contamination, respectively. However, the remaining metals exhibit practically un-contamination in the sediments. During winter, Cd indicates strong to extreme pollution; Zn causes unpolluted to moderate pollution, whereas Pb shows least to moderate contamination.

Ecological Risk Assessment

The range and mean E_i values of the heavy metals in acid extract of the sediments during summer and winter are shown in Figure 2c. The following categorization is given by Hakanson ([1980]) for E_i: $E_i < 40$ demonstrates low risk, $E_i = 40$ to 80 moderate risk, $E_i = 80$ to 160 considerable risk, $E_i = 160$ to 320 great risk, and $E_i > 320$ demonstrates very great risk. The categorization related to RI is also suggested by Hakanson ([1980]): RI < 65 explicates low risk, RI = 65 to 130 moderate risk; RI = 130 to 260 considerable risk, and RI > 260 explicates very high risk. The results elucidate that Cd causes low to very high risk, while the rest of the metals explicate low risk in the sediments during both seasons. Overall, the cumulative potential risk index (RI = 45.91 to 935 during summer and RI = 31.87 to 1,058 during winter) reveals low to very high risk of the sediments during both seasons. However, relatively higher potential ecological risk is observed during winter compared to summer.

Source Apportionment

One of the important aspect of the present study is the source apportionment of the metals in sediments using PCA and CA. The principal component loadings of the heavy metals in acid extract of the sediments during summer and winter are given in Table 4, whereas the corresponding CA is shown in Figure 3. During summer, two PCs are extracted with eigenvalues more than 1, explaining about 60% of the total variance. The first PC (36.14% variance) reveals elevated loadings of Fe, Mn, and Cu, supported by their mutual cluster in CA. These metals are likely to be contributed by lithogenic processes such as soil erosion and rock weathering. The second PC (23.77% variance) shows significant loadings of Pb, Cd, Cr, and Zn supported by their shared cluster and are mainly contributed by automobile emissions, agricultural runoff, and untreated urban wastes. The counterpart data during winter also yield two PCs with eigenvalues greater than 1, explaining more than 66% of the total variance. PC1 (51.09% variance) exhibits higher loadings for Zn, Cu, Cr, Mn, Pb, and Cd, which are predominantly contributed by transportation activities, untreated urban wastes, and agricultural runoff. The cluster analysis also shows a joint cluster for these metals. PC2 (15.22% variance) reveals the natural/

lithogenic contribution as manifested by the elevated loadings of Fe only which shows almost independent pattern in CA.

Table 4: Principal component loadings of heavy metals in acid extract of sediments during summer and winter

	Summer		Winter	
	PC1	PC2	**PC1**	**PC2**
Eigenvalue	2.530	1.664	3.576	1.065
Percentage of total variance	36.14	23.77	51.09	15.22
Percentage of cumulative variance	36.14	59.91	51.09	66.30
Cd	0.131	0.733	0.506	−0.139
Cr	0.284	0.666	0.838	−0.015
Cu	0.914	0.144	0.884	0.179
Fe	0.800	−0.108	−0.035	0.951
Mn	0.913	0.020	0.835	0.140
Pb	−0.179	0.714	0.557	0.388
Zn	−0.273	0.393	0.876	0.011

Iqbal and Shah

Iqbal and Shah Journal of Analytical Science and Technology 2014 5:28   doi:10.1186/s40543-014-0028-z

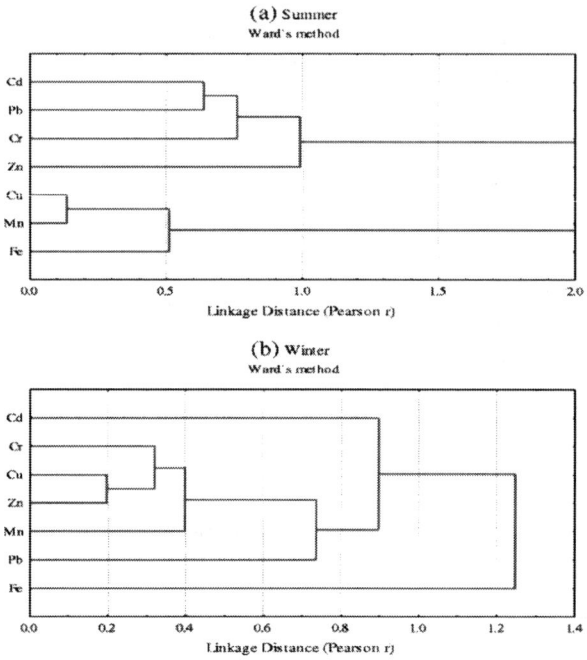

Figure 3: Cluster analyses of heavy metals in acid extract of sediments during(a)summer and (b) winter.

Sediment Quality Guidelines

The assessment of acid-extractable metal levels in the sediments is the first step to gauge the pollution of the water reservoir. However, it does not provide information on the potential toxicity to the benthic flora and fauna in the reservoir. For this purpose, numerous sediment quality guidelines are used to protect aquatic biota from the harmful and toxic effects related with sediment-bound contaminants (Caeiro et al. [2005]; McCready et al. [2006]; Spencer and Macleod [2002]). These guidelines evaluate the degree to which the sediment-associated chemical status might adversely affect the aquatic organisms and therefore are designed for the interpretation of sediment quality. SQGs have been developed for both freshwater and marine ecosystems to represent threshold chemical concentrations associated with the presence or absence of biological effects on communities (Caeiro et al. [2005]; Long and MacDonald [1998]; MacDonald et al. [2000];

Thompson et al.[2005]; Wenning et al. [2005]). These guidelines have been widely used to screen sediment contamination by comparing the concentrations in sediments with the corresponding quality guidelines in aquatic ecosystems (Caeiro et al. [2005]; MacDonald et al. [2000]). It is important to determine whether the estimated concentrations of heavy metals in sediments pose a threat to aquatic life, and they are assessed by two sets of sediment quality guidelines: (i) lowest effect level (LEL) and severe effect level (SEL) and (ii) effects range low (ERL) and effects range medium (ERM) (MacDonald et al. [2000]). These two sets of numerical SQGs are directly applied to assess the possible risk associated with heavy metal contamination in the sediments. It is interpreted that LEL and ERL as the concentrations below which adverse biological effects rarely occur. Hence, these are considered to provide a high level of protection for aquatic organisms. Similarly, SEL and ERM refer to the concentrations above which adverse biological effects frequently occur. Hence, these are considered to provide a lower level of protection for aquatic organisms (Long and MacDonald [1998]; MacDonald et al. [2000]).

The description of SQGs and sediment classification along with the results related to the sediments from Khanpur Lake during summer and winter is presented in Table 5, while the percent contribution of heavy metals towards potential acute toxicity in the sediments during summer and winter is depicted in Figure 4. During summer, the measured levels of Cd, Cr, Cu, Mn, and Pb are found to be higher than the LEL values in 87%, 100%, 100%, 37%, and 37% sediment samples, respectively. It depicts that these metals could pose moderate impact on the biota (Graney and Eriksen [2004]). On the other hand, the concentrations of Fe and Zn are found to be lower than the LEL levels in 100% sediment samples, demonstrating that these metals cause little or no impact on biota in the lake. Similarly, the measured levels of Cd, Cr, Cu, and Zn are found to be lower than the ERL values in 100% sediment samples, revealing that these metals are not associated with adverse health effects to the dwelling biota (MacDonald et al. [2000]). However, Pb levels are found to be higher in 37% sediment samples, manifesting that Pb is associated with frequent adverse biological effects to the underlying organisms (MacDonald et al. [2000]). Furthermore, potential acute toxicity ($\sum$TUs) study shows that the mean levels of toxic units (TUs) for heavy metals follow a decreasing order: Cd > Cr > Pb > Zn > Cu. It indicates relatively higher contributions of Cd, Cr, and Pb to $\sum$TUs (i.e.,

31%, 22%, and 21%, respectively; Figure 4) (Pedersen et al. [1998]). Nevertheless, Cu (11%) is the minor contributor to $\sum$TUs compared with the other heavy metals. The levels of $\sum$TUs range from 0.64 to 3.45 with a mean value of 1.75 in the sediments. Based on the USEPA sediments classification (Giesy and Hoke [1990]), Cr, Cu, and Zn show moderate contamination, Mn and Pb exhibit heavy pollution, and Cd and Fe reveal little or no contamination in the sediments during summer. It demonstrates that Cr, Cu, Zn, Pb, and Mn are the major contributors toward the gross pollution of the water reservoir.

Table 5: Description of sediment classification and sediment quality guidelines in acid extract of sediments in two seasons

		Cd	Cr	Cu	Fe	Mn	Pb	Zn
Sediment classification	Non-polluted	-	<25	<25	<17,000	<300	<40	<90
	Moderately polluted	-	25 to 75	25 to 50	17,000 to 25,000	300 to 500	40 to 60	90 to 200
	Heavily polluted	>6	>75	>50	>25,000	>500	>60	>200
Sediment quality guidelines (SQGs)	LEL	0.6	26	16	20,000	460	31	120
	SEL	10	110	110	40,000	1,100	250	820
	ERL	5	80	70	-	-	35	120
	ERM	9	145	390	-	-	110	270
Percentage of samples (summer)	Non-polluted	100	-	-	100	23	63	67
	Moderately polluted	-	100	100	-	43	23	33
	Heavily polluted	-	-	-	-	34	14	-
	<LEL	13	-	-	100	63	63	100
	≥LEL and <SEL	87	100	100	-	37	37	-
	>SEL	-	-	-	-	-	-	-
	<ERL	100	100	100	-	-	63	100
	≥ERL and <ERM	-	-	-	-	-	37	-
	>ERM	-	-	-	-	-	-	-

Percentage of samples (winter)	Non-polluted	100	3.0	40	100	53	100	97
	Moderately polluted	-	97	57	-	44	-	3.0
	Heavily polluted	-	-	3.0	-	3.0	-	
	<LEL	10	3.0		100	90	93	100
	≥LEL and <SEL	90	97	100	-	10	7.0	-
	>SEL	-	-		-	-	-	-
	<ERL	97	100	100	-	-	93	100
	≥ERL and <ERM	3.0	-	-	-	-	7.0	-
	>ERM	-	-	-	-	-	-	-

The units of metals are expressed in milligrams per kilogram. LEL, lowest effect level; SEL, severe effect level; ERL, effect range low; ERM, effect range median.

Iqbal and Shah

Iqbal and Shah Journal of Analytical Science and Technology 2014 5:28   doi:10.1186/s40543-014-0028-z

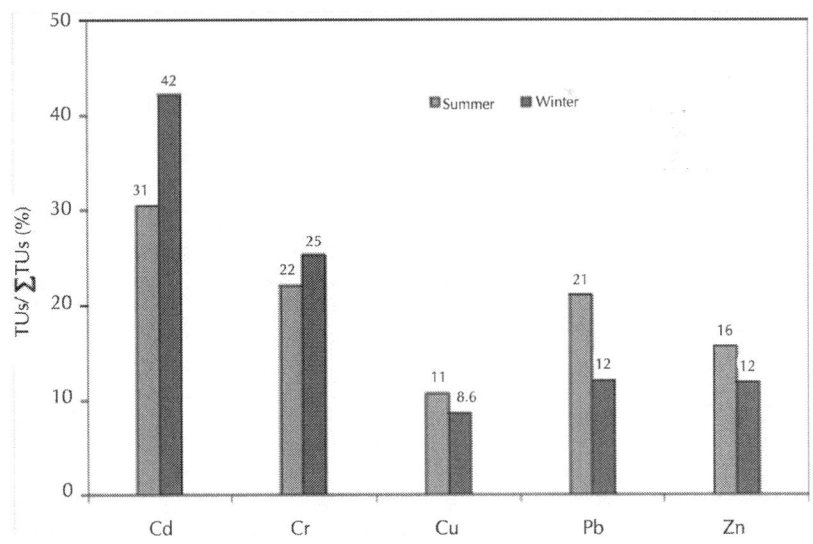

Figure 4: Percent contribution of heavy metals to ∑TUs in acid extract of sediments during summer and winter.

During winter, the measured levels of Cd, Cr, Cu, Mn, and Pb in the sediments are found to be higher than the LEL values in 90%, 97%, 100%, 10%, and 7.0% samples, respectively. It reveals moderate impact on the biota health. The observed values of Fe and Zn are found to be lower than the LEL values in 100% sediment samples, indicating that these metals are not associated with adverse impact on the biota (Graney and Eriksen [2004]). The ERL and ERM SQGs manifest that Cd and Pb levels exceed the ERL values in 3.0% and 7.0% sediment samples, respectively, demonstrating that these metals are associated with occasional adverse health hazards to the surrounding biota (MacDonald et al. [2000]). The concentrations of Cr, Cu, and Zn are lower than the ERL values in 100% sediment samples, demonstrating little or no undesirable health hazards. The potential acute toxicity study reveals that the average levels of TUs for heavy metals follow a decreasing order: Cd > Cr > Pb > Zn > Cu. It illustrates that Cd, Cr, and Pb are the major contributors to $\sum$TUs (i.e., 42%, 25%, and 12%, respectively; Figure 4), while Cu (8.6%) is a minor contributor (Pedersen et al. [1998]). The values of $\sum$TUs range from 0.54 to 3.29 with an average value of 3.29 in the sediments. Based on the USEPA sediments classification (Giesy and Hoke [1990]), Cd, Fe, and Pb may pose little or no pollution. Cr and Zn cause moderate contamination, and Cu and Mn exhibit heavy pollution in the sediments. Consequently, Cr, Cu, Mn, and Zn emerge as the major pollutants in the water reservoir during winter. Overall, the SQG results lead to the conclusion that the metals, such as Cd, Cr, Cu, Mn, and Pb are of concern during both seasons. Potential acute toxicity results demonstrate that Cd, Cr, and Pb are the major toxicants, while Zn and Cu are the minor pollutants during both seasons. However, relatively higher potential acute toxicity is observed during summer than during winter.

From the ecotoxicological dataset obtained for the US Coasts, Long et al. ([1998]) have defined several classes of toxicity probability for benthic biota: m-ERM-Q < 0.1 has a 9% probability of being toxic (based on amphipod survival test), m-ERM-Q between 0.11 and 0.5 has 21% probability of toxicity, m-ERM-Q between 0.51 and 1.5 has a probability of 49% to be toxic, and m-ERM-Q > 1.50 has 76% probability of toxicity. In the present study, the m-ERM-Q values range from 0.159 to 0.408 and 0.126 to 0.337 with the average values of 0.247 and 0.200 during summer and winter, respectively. Consequently, the metals pose approximately 21% probability of toxicity to the benthic organisms in the lake during both seasons.

Bioavailability of Heavy Metals in the Sediments

Potential toxicity of heavy metals in the sediments is also assessed by the measurement of mobile metal concentrations. The statistical distribution parameters related to the concentrations of heavy metals in $Ca(NO_3)_2$ extract of the sediments during summer and winter are given in Table 2, whereas their percent extraction in $Ca(NO_3)_2$ extract is shown in Figure 5. The summer results reveal Fe as having the highest contributions (2.069 mg/kg), while the measured levels of Mn and Cd are the least. Nonetheless, the winter results demonstrate an elevated concentration of Pb (1.205 mg/kg), and the mean level of Mn is the lowest. On the percent extraction basis, the metals follow identical decreasing sequence during summer and winter: Pb > Cd > Cr > Cu > Zn > Fe > Mn. Moreover, there are no direct relationships among the $Ca(NO_3)_2$-extractable and acid-extractable fractions of the metals in sediments. The $Ca(NO_3)_2$-extractable recoveries are found to be within approximately 11% during summer and 15% during winter of the acid-extractable metal concentrations (Figure 5). Since element bioavailability is related to its solubility, extractable metal concentrations may correspond to the bioavailable concentrations (An and Kampbell [2003]). The results demonstrate that Pb and Cd show the maximum extraction efficiencies, mobilities, and bioavailabilities, followed by Cr, while Fe and Mn manifest the least during both seasons. Accordingly, Pb, Cd, and Cr exhibit higher mobility and higher potential toxicity to the surrounding biota, while Fe and Mn show least mobility and bioavailability to the benthic biota in the water reservoir.

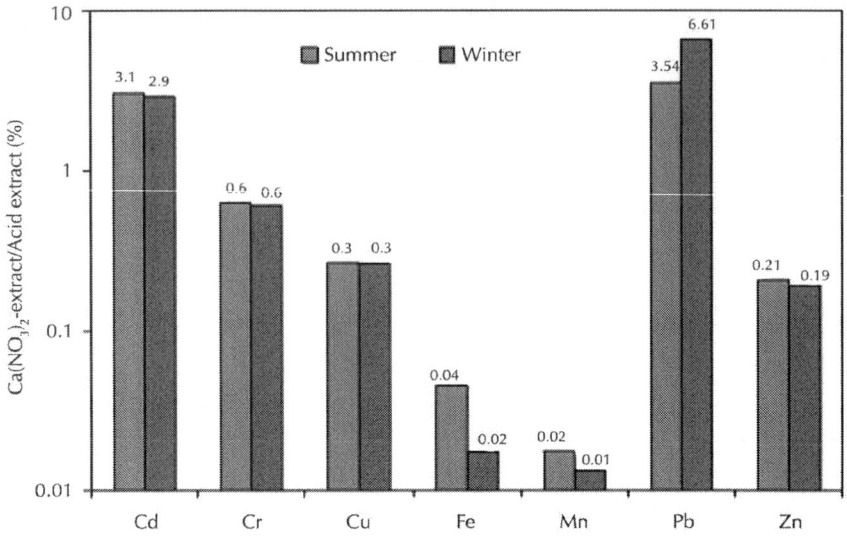

Figure 5: Percent extraction of heavy metals in $Ca(NO_3)_2$ extract of sediments during summer and winter.

CONCLUSIONS

The present study is primarily related to the evaluation of the distribution, correlation, source apportionment, contamination, and risk assessment of the heavy metals in surface sediments from Khanpur Lake, Pakistan. The study shows significantly divergent metal levels for most of the cases in the sediments during summer and winter. Most of the metals exhibit random distribution and diverse correlations in the sediments. Extremely severe enrichment is noted for Cd and Pb, while Zn shows severe enrichment. Moderate pollution is associated with Pb levels; strong to extreme pollution is shown by Cd, which is also associated with very high risk. On the whole, RI shows low to very high risk of contamination in the sediments. Multivariate PCA and CA manifest dominantly anthropogenic contributions of Pb, Cd, Cr, and Zn in the sediments. Comparison of heavy metal contents in the sediments with quality guidelines indicates adverse biological effects to the surrounding flora and fauna due to elevated levels of the metals. The m-ERM-Q study reveals 21% probability of toxicity due to the metals in the sediments. The potential toxicity, mobility, and bioavailability manifest that Cd and Pb are more mobile and available to the benthic flora and fauna. The present investigation clearly indicates that the sediments from freshwater reservoir are contaminated with some toxic heavy metals. Consequently, there is a dire need to reduce/regulate the anthropogenic sources of pollution in the study area.

AUTHORS' CONTRIBUTIONS

JI performed the field sample collection, extraction/analysis of the metals, and prepared the main draft of the manuscript. MHS designed/supervised the work, performed the statistical analysis, and helped in writing the manuscript. All authors read and approved the final manuscript.

ACKNOWLEDGEMENTS

The research fellowship awarded by Quaid-i-Azam University, Islamabad, to carry out this project is appreciatively accredited. We are also grateful to the administration of Khanpur Lake, Islamabad, for their assistance and help during the sampling campaign.

REFERENCES

1.	Adaikpoh EO, Nwajei GE, Ogala JE (2005) Heavy metals concentrations in coal and sediments from River Ekulu in Enugu, coal city of Nigeria. J Appl Sci Environ Manage 9:5-8

2.	Akoto O, Bruce TN, Darko G (2008) Heavy metals pollution profiles in streams serving the Owabi reservoir. Afr J Environ Sci Technol 2:354-359

3.	An YJ, Kampbell DH (2003) Total, dissolved, and bioavailable metals at Lake Texoma marinas. Environ Pollut 122:253-259

4.	Arnason JG, Fletcher BA (2003) A 40+ year record of Cd, Hg, Pb, and U deposition in sediments of Patroon Reservoir, Albany County, NY, USA. Environ Pollut 123:383-391

5.	Birch G, et al. (2003) A scheme for assessing human impacts on coastal aquatic environments using sediments. In: Coastal GIS.

Wollongong University Papers in Center for Maritime Policy, Australia. p 14

6. Caeiro S, Costa MH, Ramos TB, Fernandez F, Silveira N, Coimbra A, Medeiros G, Painho M (2005) Assessing heavy metal contamination in Sado estuary sediment: an index analysis approach. Ecol Indicat 5:151-169

7. Cukrov N, Bilinski SF, Hlaca B, Barisic D (2011) A recent history of metal accumulation in the sediments of Rijeka harbor, Adriatic Sea, Croatia. Mar Pollut Bull 62:154-167

8. Cuong DT, Obbard JP (2006) Metal speciation in coastal marine sediments from Singapore using a modified BCR-sequential extraction procedure. Appl Geochem 21:1335-1346

9. de Vallejuelo SFO, Arana G, de Diego A, Madariaga JM (2010) Risk assessment of trace elements in sediments: the case of the estuary of the Nerbioi–Ibaizabal River (Basque Country). J Hazard Mater 181:565-573

10. Fadigas JC, dos Santos AMP, de Jesus RM, Lima DC, Fragoso WD, David JM, Ferreira SLC (2010) Use of multivariate analysis techniques for the characterization of analytical results for the determination of the mineral composition of kale. Microchem J 96:352-356

11. Gielar A, Rybicka EH, Moller S, Einax JW (2012) Multivariate analysis of sediment data from the upper and middle Odra River (Poland). Appl Geochem 27:1540-1545

12. Giesy JP, Hoke RA (1990) Freshwater sediment quality criteria: toxicity bioassessment. In: Baudo R, Giesy JP, Muntao M (eds) Sediment: chemistry and toxicity of in-place pollutants, Lewis Publishers, Ann Arbor, MI. p 39

13. Graney JR, Eriksen TM (2004) Metals in pond sediments as archives of anthropogenic activities: a study in response to health concerns. Appl Geochem 19:1177-1188

14. Hakanson L (1980) An ecological risk index for aquatic pollution control: a sedimentological approach. Water Res 14:975-1001

15. Iqbal J, Shah MH (2011) Distribution, correlation and risk assessment of selected metals in urban soils from Islamabad, Pakistan. J Hazard Mater 192:887-898

16. Larrose A, Coynel A, Schafer J, Blanc G, Masse L, Maneux E (2010) Assessing the current state of the Gironde Estuary by mapping priority contaminant distribution and risk potential in surface sediment. Appl Geochem 25:1912-1923

17. Lide DR (2005) CRC handbook of Chemistry and Physics, Geophysics, Astronomy, and Acoustics. Section 14, Abundance of elements in the Earth's crust and in the sea. CRC Press, Boca Raton, FL.

18. Long ER, MacDonald DD (1998) Recommended uses of empirically derived, sediment quality guidelines for marine and estuarine ecosystems. Hum Ecol Risk Assess 4:1019-1039

19. Long ER, Field LJ, McDonald DD (1998) Predicting toxicity in marine sediments with numerical sediment quality guidelines. Environ Toxicol Chem 17:714-727

20. Loska K, Wiechula D, Korus I (2004) Metal contamination of farming soils affected by industry. Environ Int 30:159-165

21. Luoma SN, Rainbow PS (2008) Metal contamination in aquatic environments: science and lateral management. Cambridge University Press, Cambridge, UK.

22. MacDonald DD, Ingersoll CG, Berger TA (2000) Development and evaluation of consensus-based sediment quality guidelines for freshwater ecosystems. Arch Environ Contam Toxicol 39:20-31

23. Martin CW (2000) Heavy metal trends in floodplain sediments and valley fill, River Lahn, Germany. Catena 39:53-68

24. Martin J, Cabeza JAS, Eriksson M, Levy I, Miquel JC (2009) Recent accumulation of trace metals in sediments at the DYFAMED site (Northwestern Mediterranean Sea). Mar Pollut Bull 59:146-153

25. McCready S, Birch GF, Long ER (2006) Metallic and organic contaminants in sediments of Sydney Harbour, Australia and vicinity—a chemical dataset for evaluating sediment quality guidelines. Environ Int 32:455-465

26. Muller G (1969) Index of geoaccumulation in sediments of the Rhine River. J Geol 2:108-118

27. Nouri J, Mahvi AH, Jahed GR, Babaei AA (2008) Regional distribution pattern of groundwater heavy metals resulting from agricultural activities. Environ Geol 55:1337-1343

28. Pedersen F, Sjobrnestad E, Andersen HV, Kjolholt J, Poll C (1998) Characterization of sediments from Copenhagen harbour by use of biotests. Water Sci Technol 37:233-240

29. Radojevic M, Bashkin VN (1999) Practical environmental analysis. The Royal Society of Chemistry, Cambridge, UK.

30. Reimann C, de Caritat P (2005) Distinguishing between natural and anthropogenic sources for elements in the environment: regional geochemical surveys versus enrichment factors. Sci Total Environ 337:91-107

31. Reza R, Singh G (2010) Heavy metal contamination and its indexing approach for river water. Int J Environ Sci Technol 7:785-792

32. Rodrigues SM, Henriques B, Coimbra J, da Silva EF, Pereira ME, Duarte AC (2010) Water-soluble fraction of mercury, arsenic and other potentially toxic elements in highly contaminated sediments and soils. Chemosphere 78:1301-1312

33. Shah MH, Iqbal J, Shaheen N, Khan N, Choudhary MA, Akhter G (2012) Assessment of background levels of trace metals in water and soil from a remote region of Himalaya. Environ Monit Assess 184:1243-1252

34. Sin SN, Chua H, Lo W, Ng LM (2001) Assessment of heavy metal cations in sediments of Shing Mun River, Hong Kong. Environ Int 26:297-301

35. Singh KP, Malik A, Sinha S, Singh VK, Murthy RC (2005) Estimation of source of heavy metal contamination in sediments of Gomti River (India) using principal component analysis. Water Air Soil Pollut 166:321-341

36. Spencer KL, Macleod CL (2002) Distribution and partitioning of heavy metals in estuarine sediment cores and implications for the use of sediment quality standards. Hydrol Earth Syst Sci 6:989-998

37. Sprovieri M, Feo ML, Prevedello L, Manta DS, Sammartino S, Tamburrino S, Marsella E (2007) Heavy metals, polycyclic aromatic hydrocarbons and polychlorinated biphenyls in surface sediments of the Naples harbor (southern Italy). Chemosphere 67:998-1009

38. Tessier E, Garnier C, Mullot JU, Lenoble V, Arnaud M, Raynaud M, Mounier S (2011) Study of the spatial and historical distribution of sediment inorganic contamination in the Toulon bay (France). Mar Pollut Bull 62:2075-2086

39. Thompson PA, Kurias J, Mihok S (2005) Derivation and use of sediment quality guidelines for ecological risk assessment of metals and radionuclides released to the environment from uranium mining and milling activities in Canada. Environ Monit Assess 110:71-85

40. (2007) Microwave assisted acid digestion of sediments, sludges, soils, and oils. Method 3051A. United States Environmental Protection Agency, Office of Solid Waste and Emergency Response, US Government Printing Office, Washington, DC.

41. Varol M, Sen B (2012) Assessment of nutrient and heavy metal contamination in surface water and sediments of the upper Tigris River, Turkey. Catena 92:1-10

42. Wenning R, Batley G, Ingersoll C, Moore D (2005) Use of sediment quality guidelines and related tools for the assessment of contaminated sediments. Society of Environmental Toxicology and Chemistry (SETAC) Press, USA.

43. Yu R, Hu G, Wang L (2010) Speciation and ecological risk of heavy metals in intertidal sediments of Quanzhou Bay, China. Environ Monit Assess 163:241-252

Prediction of Heavy Metal Removal by Different Liner Materials from Landfill Leachate: Modeling of Experimental Results Using Artificial Intelligence Technique

Nurdan Gamze Turan[1], Emine Beril Gümüşel[1], and Okan Ozgonenel[2]

[1]Department of Environmental Engineering, Engineering Faculty, Ondokuz Mays University, Kurupelit, 55139 Samsun, Turkey

[2]Department of Electric and Electronic Engineering, Engineering Faculty, Ondokuz Mays University, Kurupelit, 55139 Samsun, Turkey

ABSTRACT

An intensive study has been made to see the performance of the different liner materials with bentonite on the removal efficiency of Cu(II) and

Zn(II) from industrial leachate. An artificial neural network (ANN) was used to display the significant levels of the analyzed liner materials on the removal efficiency. The statistical analysis proves that the effect of natural zeolite was significant by a cubic spline model with a 99.93% removal efficiency. Optimization of liner materials was achieved by minimizing bentonite mixtures, which were costly, and maximizing Cu(II) and Zn(II) removal efficiency. The removal efficiencies were calculated as 45.07% and 48.19% for Cu(II) and Zn(II), respectively, when only bentonite was used as liner material. However, 60% of natural zeolite with 40% of bentonite combination was found to be the best for Cu(II) removal (95%), and 80% of vermiculite and pumice with 20% of bentonite combination was found to be the best for Zn(II) removal (61.24% and 65.09%). Similarly, 60% of natural zeolite with 40% of bentonite combination was found to be the best for Zn(II) removal (89.19%), and 80% of vermiculite and pumice with 20% of bentonite combination was found to be the best for Zn(II) removal (82.76% and 74.89%).

INTRODUCTION

Industrial wastes are generated in large amounts in several industries. Because of their toxicity and nonbiodegradable nature, heavy metals are of special significance [1, 2]. Industrial waste containing heavy metals is being released into the nonengineered open dumps causing detrimental effects not only on humans but also upon environment; therefore it has become imperative to develop methods for treating such wastes [3].

The sanitary landfill method for the ultimate disposal of industrial waste continues to be widely accepted and used due to its economic advantages [4, 5]. Leachate is defined as the aqueous effluent generated as a consequence of rainwater percolation through wastes, biochemical processes in waste's cells, and the inherent water content of wastes themselves [6]. When water percolates through solid wastes, both biological materials and chemical constituents are leached into solution [7, 8]. The major concern with the movement of leachate into the subsurface aquifer is the fate of the constituents found in waste [9].

In landfill, technical and geological barrier systems are employed to minimize uncontrolled emissions from the waste into the environment.

The barrier systems contain a liner material, which should have a low hydraulic conductivity and the ability to attenuate pollutants migrating through the barrier [10]. Liner materials must be developed or improved with respect to ecological and economical requirements. Moreover, these materials to prevent or control shrinkage and/or desiccation cracking need to be further investigated (the choice of suitable liner material, modifications affecting leachate quality, etc.).

Bentonite, which is typically clay, is widely used for liner material in the barrier system. It has local availability and a low hydraulic conductivity. However, leakage can result from shrinkage cracking if only bentonite is used [11]. For this reason, a suitable sand-bentonite mixture to determine the minimum percentage of bentonite necessary to fulfil the given requirements is the main task [12]. Previous studies showed that quantities higher than 15% of bentonite as an amendment in a mixture do not lead to a significant decrease in hydraulic conductivity, while strength properties and mechanical behaviour of the mixture may be adversely affected by the clay [13, 14].

The artificial neural network (ANN) is a system of data processing based on the structure of a biological neural system. The prediction with ANN is made by learning of the experimentally generated data or using validated models [15]. Because of their reliable, robust, and salient characteristics in capturing the nonlinear relationships existing between variables (multiinput/output) in complex systems, numerous applications of ANN have been successfully conducted to solve environmental problems [16–18].

In the literature, there are few studies relating to operation problems for landfilling processes based on ANNs. In the present work, heavy metal removal during landfilling of industrial waste is investigated. The effects of various liner materials, such as bentonite, natural zeolite, expanded vermiculite, and pumice on the removal of Cu(II) and Zn(II) are examined. On the basis of batch adsorption experiments, a three-layer ANN model to predict heavy metal removal efficiency of composite used as a liner material is applied in this work. Removal of heavy metal from landfilling process is optimized to determine the optimal network structure. Finally, outputs obtained from the models are compared with the experimental data, and advantages and the further developments are also discussed.

MATERIAL AND METHODS

Materials

The three natural materials and the commercially available bentonite were investigated as a liner material in this study. Among them, natural zeolitee was obtained from the Rota Mining Industry (Gördes, Manisa, Turkey); expandable vermiculite was obtained from the Fitar Agricultural Industry (Antalya, Turkey); pumice was obtained from the Soylu Mining Industry (Nevşehir, Turkey); illite was obtained from Sud Chemie Mining Industry and Trade Co. Ltd. (Ordu, Turkey); kaolinite was obtained from the Kale Mining Industry and Trade Co. Ltd. (Çanakkale, Turkey); and bentonite was obtained from the Bensan Activated Bentonite Company (Enez, Edirne, Turkey). The chemical composition of the materials is presented in Table 1. Samples were crushed and then milled resulting in small particles with a size of about 0.5 mm.

Table 1: Chemical compositions of the liner materials

Components	B	NZ	EV	P
Na2O	1.80	0.40	0.05	3.65
MgO	4.00	1.40	17.75	0.03
Al2O3	17.00	11.80	18.45	12.27
SiO2	61.00	71.00	41.29	73.44
CaO	2.50	3.40	0.25	0.96
TiO2	—	0.10	1.21	0.10
K2O	0.50	2.40	7.21	4.37
Fe2O3	3.00	1.70	6.51	1.2
MnO	—	—	0.04	0.06
SO3	—	0.12	—	0.08
LOI	3.51	6.87	5.02	3.72
CEC (meq/100)	31.8	166.3	52.9	34.6

B: bentonite, NZ: natural zeolite, EV: expanded vermiculite, P: pumice.

Experimental Procedure

The industrial waste obtained from an electroplating industry in Samsun (Turkey) was used in the experiments. Ten simulated landfill systems were used for the removal of heavy metal from leachate. The systems were composed with a capacity of 25 L (20 cm × 25 cm × 50 cm).

Natural material-bentonite mixtures were prepared to evaluate how much these clays reduce the removal of heavy metal as liner materials. The amounts of natural materials used in the mixtures were 25%, 50%, and 75% of mixtures as volumetric. The mixture of natural liner materials is placed at the base of the simulated landfill systems. Total volume of liner material was 5 L for all systems. The system containing 100% of bentonite as a liner material was compared to other systems. A 20 L of industrial waste was deposited on the liner materials.

Leachate was collected from a drainage channel at the bottom of the system by adding distilled water. For each week, a total volume of 1 L influent was passed through the industrial waste sample. 500 mL of effluent was collected and acidified with concentrated nitric acid. The tests were conducted for 15 weeks. For each sample, the effluent was then analyzed for the heavy metal ions Cu^{2+} and Zn^{2+} using an atomic adsorption spectrophotometer (UNICAM 929 Model). Duplicate samples were prepared for all tests.

Artificial Neural Network Application

An effective way to modeling batch adsorption system for Cu(II) and Zn(II) removal is achieved by the use of artificial neural network (ANN). Nowadays, considerable achievements in artificial intelligence techniques can be used to model and predict the responses in complex systems. These techniques can enhance the predicting ability of the model such as adsorption systems if the mathematical or statistical methods fail to formulate with the desired accuracy.

A lot of scientists present ANN techniques for modeling batch experimental systems. Generally, feed-forward back propagation (FFBP) ANNs were successfully used in adsorption studies [19–24]. Details about ANNs can be found in the related literature. All these techniques use more or less the same network architecture. The optimum network type is found by trial and error, and training procedures for these suggested ANNs need long computer runs. FFBP consists of one input layer, one or several hidden layers, and one output layer. Back propagation (BP) learning algorithm is usually used for learning procedure. The mathematical background of BP algorithm can be found in [25, 26].

In this paper a simple ANN topology with 2 neurons in input layer, 2 neurons in hidden layer, and 2 neurons in output layer is needed to model the system. There are a number of common activation functions in use with ANNs. The most common choice of activation functions for multilayered perceptron (MLP) is used as hyperbolic tangent function (Figure 1).

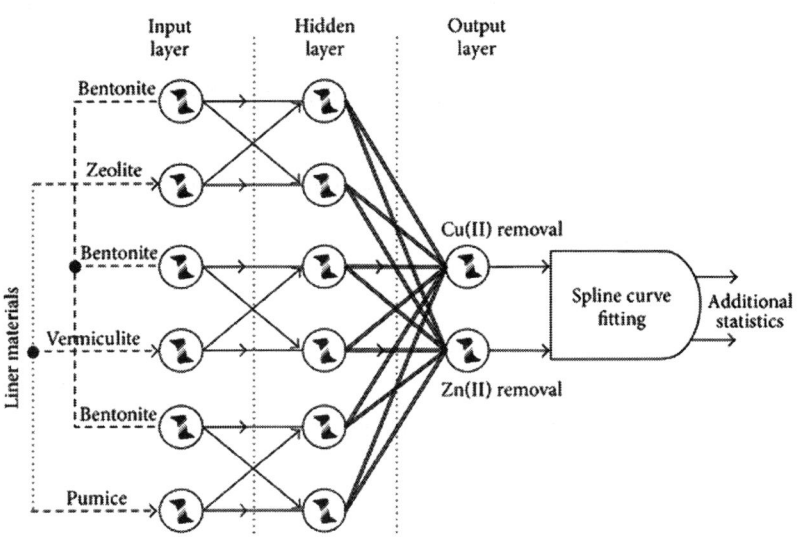

Figure 1: Proposed ANN structure for modeling adsorption system.

As seen in Figure 1 three liner materials were tested to see the efficiency of the proposed batch experimental system to maximize the Cu(II) and Zn(II) removal. Training process consists of four steps: (a) assemble the training data, (b) decide the network type, (c) train the network, and (d) calculate the output for test data. Unlike experimental design, the proposed ANN uses only 2 inputs and responses of 2, that is, Cu(II) and Zn(II) removal. Therefore, it is easy to implement and cost-effective.

The input and target data were selected from Table 2. The percentage of each liner material, that is, natural zeolite, expanded vermiculite, and pumice, and removal efficiencies were used for training procedures.

Table 2: The whole experimental system

Liner materials %	Cu (II) removal (%)	Zn (II) removal (%)
Bentonite	45.07	48.19
25% natural zeolite + 75% bentonite	63.27	60.47
50% natural zeolite + 50% bentonite	89.20	80.90
75% natural zeolite + 25% bentonite	98.34	95.29
25% vermiculite + 75% bentonite	47.97	50.31
50% vermiculite + 50% bentonite	52.73	68.98
75% vermiculite + 25% bentonite	77.41	81.49
25% pumice + 75% bentonite	52.38	50.00
50% pumice + 50% bentonite	57.50	67.50
75% pumice + 25% bentonite	78.62	74.24

Table 2 actually was used for training the ANNs according to the network parameters given in Table 3.

Table 3: Training parameters for all ANNs

Max. iteration	20000
Learn rate start control iteration	1.000
Learn rate	0.075
Min. learn rate	0.001
Max. learn rate	0.075
Momentum	0.800
Tolerance	0.000
RMS error	0.000

For testing the ANNs (Figure 1) the ratio of each liner with respect to bentonite was changed from 5% to 95% and the outputs of ANNs were predicted. This testing procedure yielded to 19 trials. Three of the total trials such as 25%, 50%, and 75% of the liner materials and 75%, 50%, and 25% of bentonite were already used in testing process and these three trials were then used to check the consistency of the ANNs. Figures 2and 3 demonstrate the performance of the suggested ANNs topology for Cu(II) and Zn(II) removal, respectively.

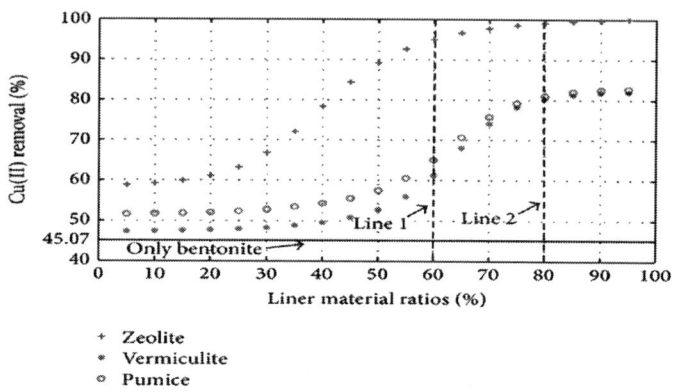

Figure 2: Prediction of Cu (II) removal with different liner materials.

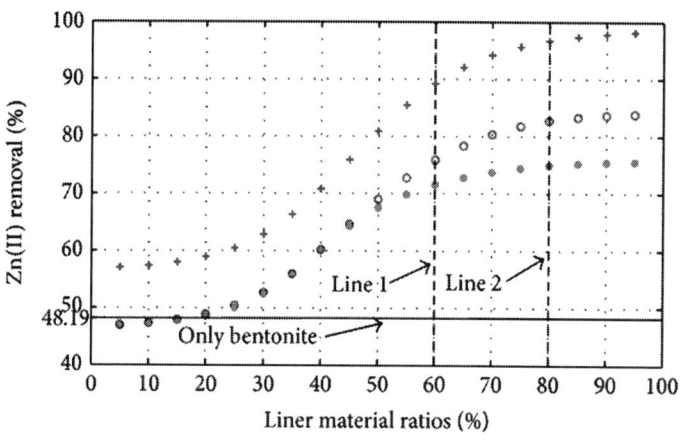

Figure 3: Prediction of Zn (II) removal with different liner materials.

In Figures 2 and 3 line 1 demonstrates the optimal point for natural zeolite material while line shows the optimal points for the materials of expanded vermiculite and pumice. Consequently, 60% of natural zeolite with 40% of bentonite combination was found to be best for Cu(II) removal (95%) and 80% of expanded vermiculite and pumice with 20% of bentonite combination was found to be the best for Zn(II) removal (61.24% and 65.09%). Similarly, 60% of natural zeolite with 40% of bentonite combination was found to be the best for Zn(II) removal (89.19%), and 80% of expanded vermiculite and pumice with 20% of bentonite combination was found to be the best for Zn(II) removal (82.76% and 74.89%). A minimum liner material and bentonite combination can be selected as 20% according to Figures 2 and 3 since there is no significant change in the removal efficiency up to that point.

The intermediate values of the outputs of ANNs can also be tested by cubic spline interpolation technique. Equation (1) gives the mathematical explanation of the interpolation technique:

$$S(z, v, p) = \begin{cases} ax^3 + bx^2 + cx + d, \\ ex^3 + fx^2 + gx + h. \end{cases}$$

(1)

In (1) z, v, and p stand for natural zeolite, expanded vermiculite, and pumice, respectively, and x represents the interval areas, that is, from 5% to 95%. Table 4 gives the descriptive statistics of the interpolated outputs of ANNs.

Table 4: Basic statistics for interpolated ANN outputs

Descriptive statistics	Cu(II) natural zeolite	Cu(II) vermiculite	Cu(II) pumice	Zn(II) natural zeolite	Zn(II) vermiculite	Zn(II) pumice
Minimum	58.83	47.37	51.63	57.06	46.57	46.57
Maximum	99.93	82.05	82.77	98.15	83.89	75.52
Mean	82.74	60.60	63.86	78.69	66.68	63.38
Median	89.20	52.73	57.50	80.90	68.98	67.50
Mod	58.83	47.37	51.63	57.06	46.97	46.97

Standard value	16.62	14.35	12.74	16.38	14.64	11.46
Range	41.10	34.68	31.13	41.09	36.93	28.55

The removal efficiencies were calculated as 45.07% and 48.19% for Cu(II) and Zn(II), respectively (Table 1), when only bentonite was used as liner material. However, the use of other liner materials in specific ratios had significant effect on removal efficiencies (Table 3).

CONCLUSIONS

The idea of the study was to examine the feasibility of using different liner materials to remove Cu(II) and Zn(II) from industrial leachate. The following outcomes can be derived from this ongoing research work.

*   The traditional use of bentonite as a liner material has low removal efficiency comparing to combinations of natural zeolite + bentonite, expanded vermiculite + bentonite, and pumice + bentonite mixtures.

*   The suggested ANN topology was found effective to model the experimental design.

*   Applying cubic spline curve fitting of the removal efficiencies enables the provision of additional descriptive statistics.

*   Among the different combinations of liner materials 60% of natural zeolite + 40% of bentonite was found the optimum with high removal efficiencies of 95% for Cu(II) and 89.19% for Zn(II).

REFERENCES

1.   M. Alkan, B. Kalay, M. Do⊏an, and Ö. Demirbaş, "Removal of copper ions from aqueous solutions by kaolinite and batch design," Journal of Hazardous Materials, vol. 153, no. 1-2, pp. 867–876, 2008.

2.   F. Claret, C. Tournassat, C. Crouzet et al., "Metal speciation in landfill leachates with a focus on the influence of organic matter," Waste Management, vol. 31, no. 9-10, pp. 2036–2045, 2011.

3.   T. Aman, A. A. Kazi, M. U. Sabri, and Q. Bano, "Potato peels as solid waste for the removal of heavy metal copper(II) from waste water/industrial effluent," Colloids and Surfaces B, vol. 63, no. 1, pp. 116–121, 2008.

4.   P. H. Brunner and J. Fellner, "Setting priorities for waste management strategies in developing countries," Waste Management and Research, vol. 25, no. 3, pp. 234–240, 2007.

5.   D. Laner, M. Crest, H. Scharff, J. W. F. Morris, and M. A. Barlaz, "A review of approaches for the long-term management of municipal solid waste landfills," Waste Management, vol. 32, no. 3, pp. 498–512, 2012.

6.   S. Renou, J. G. Givaudan, S. Poulain, F. Dirassouyan, and P. Moulin, "Landfill leachate treatment: review and opportunity," Journal of Hazardous Materials, vol. 150, no. 3, pp. 468–493, 2008.

7.   T. Chalermyanont, S. Arrykul, and N. Charoenthaisong, "Potential use of lateritic and marine clay soils as landfill liners to retain heavy metals," Waste Management, vol. 29, no. 1, pp. 117–127, 2009.

8.   G. di Bella, D. di Trapani, G. Mannina, and G. Viviani, "Modeling of perched leachate zone formation in municipal solid waste landfills," Waste Management, vol. 32, no. 3, pp. 456–462, 2012.

9.   N. Yusof, A. Haraguchi, M. A. Hassan, M. R. Othman, M. Wakisaka, and Y. Shirai, "Measuring organic carbon, nutrients and heavy metals in rivers receiving leachate from controlled and uncontrolled municipal solid waste (MSW) landfills," Waste Management, vol. 29, no. 10, pp. 2666–2680, 2009.

10.   K. Czurda, "Encapsulation parameters in waste deposit technology: geologic barriers and liner systems," Geo.Alp, vol. 3, pp. 207–214, 2006.

11.   L. H. Mollins, D. I. Stewart, and T. W. Cousens, "Predicting the properties of bentonite-sand mixtures," Clay Minerals, vol. 31, no. 2, pp. 243–252, 1996.

12.   T. B. Musso, K. E. Roehl, G. Pettinari, and J. M. Vallés, "Assessment of smectite-rich claystones from Northpatagonia for their use as liner materials in landfills," Applied Clay Science, vol. 48, no. 3, pp. 438–445, 2010.

13. T. C. Kenney, W. A. van Veen, M. A. Swallow, and M. A. Sungaila, "Hydraulic conductivity of compacted bentonite-sand mixtures," Canadian Geotechnical Journal, vol. 29, no. 3, pp. 364–374, 1992.

14. T. Abichou, C. H. Benson, and T. B. Edil, "Micro-structure and hydraulic conductivity of simulated sand-bentonite mixtures," Clays and Clay Minerals, vol. 50, no. 5, pp. 537–545, 2002.

15. M. R. Fagundes-Klen, L. G. L. Vaz, M. T. Veit, C. E. Borba, E. A. Silva, and A. D. Kroumov, "Biosorption of the copper and cadmium ions-a study through adsorption isotherm analysis,"Bioautomation, vol. 7, no. 1, pp. 23–33, 2007.

16. D. Salari, N. Daneshvar, F. Aghazadeh, and A. R. Khataee, "Application of artificial neural networks for modeling of the treatment of wastewater contaminated with methyl tert-butyl ether (MTBE) by UV/H_2O_2 process," Journal of Hazardous Materials, vol. 125, no. 1–3, pp. 205–210, 2005.

17. K. Yetilmezsoy and S. Demirel, "Artificial neural network (ANN) approach for modeling of Pb(II) adsorption from aqueous solution by Antep pistachio (Pistacia Vera L.) shells," Journal of Hazardous Materials, vol. 153, no. 3, pp. 1288–1300, 2008.

18. S. Aber, A. R. Amani-Ghadim, and V. Mirzajani, "Removal of Cr(VI) from polluted solutions by electrocoagulation: modeling of experimental results using artificial neural network," Journal of Hazardous Materials, vol. 171, no. 1–3, pp. 484–490, 2009.

19. N. G. Turan, B. Mesci, and O. Ozgonenel, "The use of artificial neural networks (ANN) for modeling of adsorption of Cu(II) from industrial leachate by pumice," Chemical Engineering Journal, vol. 171, no. 3, pp. 1091–1097, 2011.

20. D.-J. Choi and H. Park, "A hybrid artificial neural network as a software sensor for optimal control of a wastewater treatment process," Water Research, vol. 35, no. 16, pp. 3959–3967, 2001.

21. Ö. Çinar, H. Hasar, and C. Kinaci, "Modeling of submerged membrane bioreactor treating cheese whey wastewater by artificial neural network," Journal of Biotechnology, vol. 123, no. 2, pp. 204–209, 2006.

22. A. Wang, C. Liu, H. Han, N. Ren, and D.-J. Lee, "Modeling denitrifying sulfide removal process using artificial neural

networks," Journal of Hazardous Materials, vol. 168, no. 2-3, pp. 1274–1279, 2009.

23. R. M. Aghav, S. Kumar, and S. N. Mukherjee, "Artificial neural network modeling in competitive adsorption of phenol and resorcinol from water environment using some carbonaceous adsorbents,"Journal of Hazardous Materials, vol. 188, no. 1–3, pp. 67–77, 2011.

24. D. Saha, A. Bhowal, and S. Datta, "Artificial neural network modeling of fixed bed biosorption using radial basis approach," Heat and Mass Transfer, vol. 46, no. 4, pp. 431–436, 2010.

25. R. A. Chayjan and M. Esna-Ashari, "Comparison between artificial neural networks and mathematical models for estimating equilibrium moisture content in raisin," Agricultural Engineering International: The CIGRE Journal, vol. 12, no. 1, p. 158, 2010.

26. R. Singh, R. S. Bhoopal, and S. Kumar, "Prediction of effective thermal conductivity of moist porous materials using artificial neural network approach," Building and Environment, vol. 46, no. 12, pp. 2603–2608, 2011.

Chapter 4

Review of 15 Years of Research on Sediment Heavy Metal Contents and Sediment Nutrient Release in Inland Aquatic Ecosystems, Turkey

Serap Pulatsüand and Akasya Topçu

Department of Aquaculture and Fisheries Engineering, Agricultural Faculty, Ankara University, Ankara, Turkey

ABSTRACT

Turkey's inland water ecosystem consists of 33 rivers (177.714 miles), 200 natural lakes (906.118 ha), 159 reservoirs (342.377 ha) and 750 ponds (15.500 ha). Sedimentological studies conducted on inland water ecosystems during the last 15 years in Turkey can be categorized into two main topics. The first group of studies is concerned with heavy metal levels in sediment, with especial reference to the interaction between water, sediment and aquatic organisms. Additionally, the studies in question deal with the potential impacts of heavy metal concentrations on the ecosystem. The second group of studies is concerned with the role of eutrophication in the sediment as a result of serious contamination of inland water ecosystems. It is known that the sediment can directly influence the nutrient level in standing inland waters such as lakes and ponds by way of internal nutrient loading. In this context, studies regarding sediment, overlying water, sediment pore water and nutrient release from the sediment should be emphasized as these are important steps with respect to the eutrophication process. By keeping these studies in mind, the researcher in this study compiled and analyzed studies dealing with inland water ecosystems with differing nutrient levels and uses, including for drinking water, in Turkey's drainage basins. In addition, field and laboratory studies regarding nutrient release from sediment into Turkey's inland water ecosystems were evaluated in light of lake management practices.

RESEARCH REGARDING SEDIMENT HEAVY METAL CONTENT IN INLAND AQUATIC ECOSYSTEMS

The concentrations of heavy metals in receiving environments have both natural causes, such as sea-bed volcanic activity, atmospheric convection, rivers or erosion, and man-made causes, such as mining, the rapid increase of treatment and refining systems, the excessive use of fossil fuels, and the use of metallic products in agriculture (like arsenic pesticides). Of the heavy metals that are transported into the water, one portion is diffused in the water and the other portion forms solid compounds with carbonate, sulfate and sulfur, sinks to the bottom, and is accumulated in the sediment [1]. Accordingly, the sediment forms a trap for heavy metals, and thus, concentrations of heavy metals in the sediment can be several orders of magnitude greater than in the overlying water. Metals found in the sediment directly threaten detrital and deposit-feeding benthic organisms, and could possibly be a long-term source of contamination higher up the food chain [2] .

Fish can take in heavy metals through respiration (through the gills or the skin), adsorption to body surfaces or feeding. The intake of heavy metals and their accumulation in the organisms in aquatic ecosystems are affected by several factors, such as differences in the amount of metals entering the environment, the condition of the organism and the physical and chemical properties (temperature, salinity, pH and dissolved oxygen) of the aquatic environment in which the organisms are found.

Quantitative sediment quality guidelines (SQG) exist for freshwater ecosystems, and these provide a reliable benchmark for assessing sediment quality in freshwater ecosystems. The threshold effect level (TEL) and the probable effect concentrations (PEC) for different sediment contaminants in freshwater ecosystems were determined by [3]. TEL corresponds to the concentration of a contaminant below which harmful effects on benthic organisms are expected to occur only rarely, and PEC represents the concentration above which harmful effects on aquatic biota are expected to occur frequently [3].

For this review, studies on the interaction between water, sediment and aquatic organisms conducted over the past 15 years in five different types of aquatic ecosystem in Turkey (lakes, reservoirs, wetlands, rivers and streams) were compiled, and the surface sediment concentrations of arsenic, cadmium, chromium, copper, lead, mercury, nickel and zinc reported by these studies were compared with [3]'s threshold effect level (TEL) and probable effect concentration (PEC) values for these metals. Additionally, studies performed under field and laboratory conditions regarding the release of nutrients from the sediment in Turkey's inland aquatic ecosystems were evaluated in the light of lake management practices.

Lakes

Some heavy metal concentrations in sediment from different lakes in Turkey are shown in Table 1.

Beyşehir Lake is a conservation area located in southwestern Turkey used for irrigation and drinking water. The main factors contributing to its contamination are industrialization and urbanization. During the spring of 2001, water, sediment, plankton and fish samples were collected from three different stations in the lake. The order of heavy metals accumulation in the food chain was determined as water > plankton > sediment > fish tissues, except for Cr. The highest concentration in sediment samples was recorded for Pb and the lowest for Hg. Concentrations of Cd, Pb, Hg and Cr in the lake were recorded as 13.05, 32.65, 0.24 and 10.63 µg $\cdot$ g^{-1}, respectively. According to the research, there were high concentrations of heavy metals in the water, sediment, plankton and fish. Direct contamination of the water by metals or the geochemical structure of the region was cited as possible reasons for this situation [4]. The mean concentration of mercury in the sediment was much higher than TEL, and the mean concentration of cadmium was much higher than both the TEL and PEC values given by [3].

Kovada Lake is located in the Lakes Region, Turkey's most visited wetland, which is extremely rich in terms of flora and fauna. The lake gained importance due to its surrounding forest and its natural beauty, and was given national park status and protection in 1970. In an analysis of heavy metals in the sediment of the lake, which has a maximum depth of five m, Cr, Cu, Fe, Mn, Pb, Zn, Al and Ni were present in all seasons, but Cd remained under the ICP-OES analytic limit in the summer and fall of 2005 [5] . Only the nickel concentration in the sediment was higher than TEL values.

Heavy metal concentrations from several fish species (Cyprinus carpio, Capoeta tinca, Leiciscus cephalus, Carassius gibelio and Silurus glanis) and sediment samples from six different lakes in Tokat province (Bedirkale, Boztepe, Belpınarı, Avara, Ataköy and Akın) were determined by [6] in the spring and summer of

Table 1: Heavy metal concentrations (mg · g^{-1} DW) in sediment samples from different lakes in Turkey

Heavy metal conc. / Lakes	As	Cd	Cr	Cu	Pb	Hg	Ni	Zn	References
Beyşehir Lake	-	13.05[1,2]	10.63	-	32.65	0.241	-	-	[4]
Kovada Lake	-	nd-0.27	6.63 - 17.59	4.65 - 13.77	1.96 - 4.42	-	9.13 - 25.931	12.82 - 33.42	[5]
(Tokat Lakes) Bedirkale	-	-	4.5 - 6.1	5.1 - 8.2	2.9 - 3.4	-	51.1 - 53.61,2	33.7 - 38.9	[6]
Boztepe	-	-	6.4 - 7.8	4.7 - 6.1	6.7 - 7.0	-	51.6 - 55.41,2	23.3 - 23.9	
Belpınarı	-	-	9.4 - 10.7	5.8 - 6.3	3.9 - 4.0	-	37.8 - 38.01	26.5 - 29.7	
Avara	-	-	5.1 - 5.6	3.4 - 3.7	3.8 - 5.1	-	42.1 - 42.71	30.4 - 32.3	
Ataköy	-	-	4.7 - 4.9	6.3 - 6.8	2.7 - 3.1	-	50.6 - 55.01,2	24.7 - 25.0	
Akın	-	-	4.4 - 4.6	4.6 - 5.0	2.9 - 3.5	-	40.7 - 43.11	23.3 - 26.5	
Hazar Lake	-		29.0 - 35.0	24.0 - 51.01(H)	nd	-	41.01	49.0 - 70.0	[7]
	-	nd	87.70 - 43.341	46.32 ± 9.141	1.14 ± 3.49		49.50 ± 28.611,2	27.09 ± 10.26	[8]
Terkos Lake		0.56 - 1.161	91.88 - 123.921,2(H)	5.94 - 33.09	22.61 - 62.511(H)		25.07 - 39.411	76.22 - 137.551	[9]

	As	Cd	Cr	Cu	Pb	Hg	Ni	Zn	Reference
Uluabat Lake	-	0.02 - 0.12	0.83 - 4.88	0.25 - 1.03	0.42 - 2.39	-	1.84 - 8.93	0.74 - 8.36	[12]
	-	2.01	9.0	12.0	13.0	-	8.0	1.0	[13]
Ikli Lake	1.14 - 16.33[1]	0.691	57.91	119.21	110.71	-	209.4[1,2]	171.01	[14]
		0.09 - 3.06[1]	11.60 - 269.55[2](H)	2.68 - 38.84[1](H)			11.23 - 211.16[1,2](H)	16.34 - 159.56[1](H)	[15]
Yeniça a Lake	-	0.81	92.81	-	16	-	-	-	[16]
Mogan Lake	-	-	28.55	15.13	0.82	-	-	13.79	[17]
Karata Lake	-	0.11 - 0.25	13.82 - 53.13[1](H)	1388 - 32.27	0.54 - 1.13	-	47.16 - 203.92[1,2]	12.92 - 45.00	[18]
SQGs	As	Cd	Cr	Cu	Pb	Hg	Ni	Zn	Reference
TEL1	5.9	0.596	37.3	35.7	35.0	0.174	18.0	123	[3]
PEC2	33.0	4.98	111.0	149.0	128.0	1.06	48.6	459.0	

[1]Every superscript with 1 for heavy metal concentrations mean the concentration exceeds the TEL value; [2]Every superscript with 2 for heavy metal concentrations mean the concentration exceeds the PEC value.

2003-2004. In the fish samples, the maximum Fe, Zn, Cu, Pb, Cr, Mn and Ni concentrations were 167, 48.6, 3.6, 2.8, 1.6, 64.3 and 5.6 $\mu g \cdot g^{-1}$, respectively. As for the sediment samples, the maximum Fe, Mn, Zn, Cu, Pb, Cr and Ni concentrations of were 2138, 232, 38.9, 8.2, 7.0, 10.7 and 55.4 $\mu g \cdot g^{-1}$, respectively. The nickel values of three of the six lakes (Bedirkale, Boztepe ve Ataköy) exceeded both the TEL and the PEC values for nickel outlined by [3] .

Lake Hazar (Elazığ, Turkey) is one of the largest and deepest lakes in Eastern Anatolia. Alburnus heckeli and Aphanius asquamatus are native fish species of the lake. [7] measured the concentration of heavy metals and natural gross radioactivity in the surface water and sediment of the lake. Samples were taken from three sites during 2004 and 2005. The results showed that, in general, concentrations of heavy metals (Zn, Fe, Mn, Ni, Cu and Pb) and major elements (Na, K, Ca, Mg) in the water did not exceed the four different guidelines (WHO; World Health Organization, 1999, EC; Europe Community, 1998, EPA; Environment Protection Agency, 2002 and TSE-266; Turkish Standard, 1997). The sediment heavy metals and major elements concentration were found to decrease in the sequence Fe > Mg > Ca > Mn > Zn > Ni > Cr > Cu > Co > Pb. Of the heavy metals in the sediment (depth: 8 m), only the nickel concentrations and mean Cu highest value exceeded the highest TEL values indicated by [3] .

In an additional study of the same lake conducted by [8] heavy metal concentrations (Cd, Cr, Cu, Fe, Mn, Ni, Pb and Zn) were reported in the water, sediment, sago pondweed (Potamogeton pectinatus L.) and the muscle, livers and gills of freshwater fish (Capoeta capoeta umbla). Samples of these items were collected seasonally between 2004 and 2005 from three separate sites. The highest concentrations of metal in the water samples were found for Fe. As for sediment, the highest concentration was recorded for Mn and the lowest for Pb in all seasons and at all three sites. Accordingly, the researchers concluded that measures should be taken in Hazar Lake to prevent any further heavy metal contamination. The mean concentrations of chromium and copper in the sediment were much higher than the TEL, and the mean concentration of nickel was much higher than both the TEL and PEC values determined by [3] .

A study performed by [9] on Terkos Lake, which provides a portion of the drinking water to Turkey's most important metropolitan area,

Istanbul. In the May 2008 study, total metal accumulation (Al, Cu, Mn, Pb, Cd and Fe) in the organs and eggs of Astacus leptodactylus (Eschscholtz 1823) and in the sediment (Al, Cu, Mn, Pb, Cd, Fe, Zn, Cr, Ni)" were examined. The sediment was determined to have high enrichment factors (EF) of Zn, Cr, Cd and Pb metals, which come from anthropogenic sources (domestic and industrial input), while they have low EFs of Fe, Ni, Cu coming from natural (terrigeneous) inputs. Moreover, these lake sediments showed no Al, Fe, Ni and Cu metal enrichment due to low contamination factor (CF) values. However, the lake sediment is moderately contaminated by Zn, Cr and Pb, and heavily contaminated by Cd. The metal concentrations obtained from the sediment samples were compared with the SQGs [3] which showed that cadmium, lead (highest values), nickel and zinc concentrations exceeded TEL maximum values, and chromium exceeded both TEL and PEC maximums.

Uluabat Lake, home to Turkey's largest water lily beds and declared a Ramsar protected wetland in 1998, is located south of the Marmara Sea in Bursa province and lake has an area of between 120 and 240 km^2. The Lake Uluabat drainage is located in one of the most productive agricultural regions of Turkey. There are 16 urban settlements on the lake shores, and the main human activity on the lake is fishing. In addition to fisheries and agriculture, livestock breeding is also an important activity near the lake [10] [11] . In a study conducted by [12] on Uluabat Lake between November 2001 and September 2002, water and sediment samples taken at different stations and in various months were examined. The highest mean levels of Cu (1.19 ± 0.05 μg · g^{-1}), Cr (4.88 ± 0.30 μg · g^{-1}, Pb (2.39 ± 0.24 μg · g^{-1}), Ni (8.93 ± 1.18 μg · g^{-1}) and Cd (0.12 ± 0.006 μg · g^{-1}) were found in May 2002, in sediment samples. The highest concentration of Zn (8.36 ± 0.13 μg · g^{-1}) was recorded in September 2002. Pb concentration levels in the sediment were consistently higher than Cd concentration levels. The high levels of Pb, Cd, Zn and Fe in the sediment in the lake are a result of polluting activities around it. The heavy metal concentrations recorded by [12] did not exceed the TEL and PEC values indicated by [3] .

In another study conducted in the same lake by [13] , heavy metal accumulations (Cu, Ni, Zn, Cd, Pb, Cr, B, As) were measured in water, plankton and sediment samples taken from different areas of the lake between January 2003 and February 2004. The degrees of heavy metal concentrations in water and plankton samples were determined as B > Zn > As > Cd > Pb > Ni > Cr and Zn > Ni > Cu > Cr > Cd > Pb, respectively. The mobile heavy metals in the sediment samples were sequenced as Pb > Cu > Cr > Ni > Cd > Zn, whereas the degree of easily mobilizable metal concentration was determined as Pb > Ni > Cr > Cu > Cd > Zn. Only the cadmium concentration in the sediment exceeded TEL values reported by [3] .

A third study done on the same lake by [14] collected two replicate sediment, zoo benthic, and water samples each month between August 2004 and July 2005 from 12 sites in Lake Uluabat. The results of the study showed relatively high occurrence of metals in the water, the sediment, and the two zoo benthic taxa. The lead, chromium, and cadmium concentration levels in the sediment samples (Table 1) were higher than those reported by [13] . One possible reason for this is that cadmium and chromium in sediments are associated with carbonate fraction, and that they are easily solubilized metals. The high levels of zinc, nickel, copper, and lead observed in the study could have been due to domestic and industrial effluents, municipal runoffs, or atmospheric deposition. Thus, the main areas in need of regular review and control are discharges, pollutants, and the long-term effects of pollutants on the lake's ecosystem. According to the results of the chemical analysis of the sediment samples, the mean concentrations of cadmium, chromium, copper, lead and zinc in the sediment were much higher than the TEL values, and in the case of nickel, higher than both the TEL and the PEC values reported by [3] .

Işıklı Lake is located in the southwest of Turkey and is used for irrigation. With a depth of approximately seven m. and an area of 9.749 ha, it is fed by the Büyük Menderes, Karanlık and Kuti Rivers. There are small rush islands in the lake, and the surrounding shores are home to apple, cherry and peach orchards, grains fields, restaurants and hotels. The most common fish species in the lake are Cyprinus carpio, Esox lucius, Tinca tinca and two local fish species (Aphanius anatoliae and Chondrostoma meandrense). The lake is polluted by agricultural runoff and domestic effluents. Several heavy metals (Ba, As, Co, Cd, Cr, Cu, Fe, Mn, Ni and Zn) were detected in water, sediment and a total of 144 fish samples from Işıklı Lake between March 2009 and February 2010. Among the metals observed, Fe had the highest concentrations in both the water and the sediment, while Cd had the lowest value in the sediment, ranging between 0.19 - 3.06 mg $\cdot$ kg^{-1}. The results of this study point to a potential risk developing in the future depending on agricultural development [15] . While the concentrations of arsenic and cadmium in the sediment as well as the highest values of copper and zinc observed by the researchers were seen to be higher than the TEL values, and the highest measurement of chromium exceeded PEC values, nickel's highest observed value was greater than both the TEL and the PEC values reported by [3] .

Yeniçağa Lakein Bolu province is an extremely important aquatic ecosystem due to its position on the migration route of birds coming over the straits of Istanbul and Çanakkale, the peat beds on its perimeter, and the fishing that is practiced in the lake. It is also used for irrigation and recreational purposes. In a study conducted by [16] of the water, sediment and crayfish (Astacus leptodactylus) in the lake, some heavy metal (Cr, Cd, Mn, Pb) accumulation was observed, and it was determined that this situation had become dangerous both for the ecosystem and for public health. In the sampling done in 2004, the concentrations of Cd, Cr, Mn, and Pb in the sediment were recorded as 0.8 $\mu g \cdot g^{-1}$, 92.8 $\mu g \cdot g^{-1}$, 1143 $\mu g \cdot g^{-1}$, and 16 $\mu g \cdot g^{-1}$, respectively. The levels of cadmium and chromium in the sediment exceeded the TEL values determined by [3] .

Mogan Lake, located 20 km to the south of Turkey's capital city, Ankara, is near the town of Gölbaşı, which has undergone considerable development in recent years due to population increase and urban settlement. In a study of the lake conducted by [17] in March 2007, some metals (Al, Cr, Cu, Fe, Mn, Pb, Zn and V) and metalloid (Si) were observed in the water, sediment and some tissues of the fish Cyprinus carpio. The metal concentrations in the sediment samples were as follows: Cr: 22.19 - 41.31 $\mu g \cdot L^{-1}$, Cu: 9.91 - 30.19 $\mu g \cdot L^{-1}$, Pb: 0.46 - 1.78 $\mu g \cdot L^{-1}$, and Zn: 11.27 - 18.01 $\mu g \cdot L^{-1}$. According to the researchers, the correlation coefficient for Cu and Fe between the sediment and the common carp's (Cyprinus carpio) muscles was greater than that of the sediment and the fish's gills. Accordingly, measures should be taken to prevent metal contamination in future. The sediment metal concentrations obtained from the samples were compared with the SQGs [3] , which showed that these concentrations did not exceed the TEL and PEC levels.

Karataş Lake, in the southwest of Burdur province, is used primarily for irrigation. It also hosts migrating birds and seven fish species, one of which is endemic. A study of the water, sediment and the gills of the

Sander lucioperca in the lake done by [18] revealed some heavy metal concentrations at three sampling stations in July 2010, October 2010, January 2011 and April 2011. The heavy metal concentrations in the sediment were recorded as Fe > Mn > Ni > Zn > Cr > Cu > Se > Pb > Mo > Cd. Of the heavy metal concentrations in the sediment, chromium's highest value exceeded the TEL value, and the nickel concentration exceeded both the TEL and PEC values.

Wetlands

Some heavy metal concentrations of sediment samples from different wetlands and dam lakes in Turkey are presented in Table 2.

Akyatanis the biggest fishery lagoon in Iskenderun Bay on Turkey's Mediterranean coast. Akyatan Lagoon is a multipurpose wetland ecosystem that has important contributions to the local economy, such as aquaculture and touristic activities [11] . In a study of the lagoon carried out by [19] between December 2000 and November 2001 which looked at seasonal changes in accumulations of the heavy metals Pb, Cu, Zn, Cd, Fe in seston, benthos, and sediment samples, it was observed that heavy metal accumulation in the sediment was ordered as Fe > Pb > Zn > Cu > Cd. The seasonal changes of metal concentrations in these samples may be attributable to natural factors like the growth and reproductive cycles of the organisms and changes in the water temperature. The sediment cadmium and lead concentrations observed in the study were greater than the TEL values for these metals indicated by [3] .

Yedigöller is the nearest wetland to the Kütahya city center and provides irrigation water to some of the farmland surrounding it. Yedigöller is eutrophic, and has been used as a waste disposal area (landfill) since 1977. Of the seven lakes in the wetlands, two are now dry. [20] stress that while the Cd and Hg concentrations in the

Table 2: Heavy metal concentrations (mg · g⁻¹ DW) in sediment samples from different wetlands and dam lakes in Turkey

Heavy metal cons. Wetlands and Dam Lakes	As	Cd	Cr	Cu	Pb	Hg	Ni	Zn	References
Akyatan Lagoon	-	1.41 - 1.61[1]	-	15.87 - 22.81	35.14 - 58.40[1]	-	-	27.90 - 52.46	[19]
Yedi Göller	-	nd	43.20 - 43.93[1]	67.80 - 68.53[1]	30.67 - 32.67	nd	59.87 - 60.33,1,2	68.47 - 104.13	[20]
Seyhan Dam Lake	-	2.15 ± 0.38[1]	118.95 ± 21.71,2	19.80 ± 4.57	-	-	-	39.09 ± 6.50	[21]
Porsuk Dam Lake	-	3.36 ± 0.13[1]	78.40 ± 4.57[1]	26.08 ± 1.20	90.00 ± 40.66[1]	nd	159.12 ± 5.65,2	656.40 ± 32.391,2	[22]
Enne Dam Lake	-	5.04 ± 0.15,1,2	59.08 ± 1.79[1]	27.84 ± 0.68	88.96 ± 3.22[1]	nd	136.82 ± 5.101,2	272.00 ± 11.471	
Atatürk Dam Lake Bozyazı (Bozova)	-	nd	-	14.57	nd	nd	43.691	60.79	[23]
Akpınar (Adıyaman)	-	nd	-	22.70	nd	nd	139.691,2	59.14	

	As	Cd	Cr	Cu	Pb	Hg	Ni	Zn	Reference
Atatürk Dam Lake	-	nd	-	35.05 ± 5.02	-	nd	$72.50 \pm 28.42^{1,2}$	35.15 ± 3.32	[24]
Demirköprü Dam Lake	-	0.70 - 0.82[1]	3.58 - 6.75	9.3 - 15.1	2.66 - 6.50		7.41 - 14.30	-	[25]
Av ar Dam Lake	-	0.76[1]	13.33 - 14.48	23.47 - 29.98	2.44 - 4.04	-	28.25 - 29.99[1]	-	[26]
Geyik Dam Lake	-	nd - 2.0	-	11.0 - 120.01(H)	28.0 - 31.3	-	104.0 - 320.0^{1,2}	32.0 - 104.0	[27]
Çatören Dam Lake	-	12.4^{1,2}	-	-	-	-	-	32.0	[28]
SQGs	As	Cd	Cr	Cu	Pb	Hg	Ni	Zn	Reference
TEL1	5.9	0.596	37.3	35.7	35.0	0.174	18.0	123	[3]
PEC2	33.0	4.98	111.0	149.0	128.0	1.06	48.6	459.0	

[1]Every superscript with 1 for heavy metal concentrations mean the concentration exceeds the TEL value; [2]Every superscript with 2 for heavy metal concentrations mean the concentration exceeds the PEC value.

sediment are below measurable values, the wetland is under the pressure of extreme pollution due to its use as a landfill. Therefore, the authors of the study maintain that several measures should be taken in the area: first and foremost, leachates from the landfill should be prevented from contaminating the underground and surface water, aerobic and anaerobic bacteria should be used to break down organic matter, waste should be sorted at its source for recycling, a proper solid waste disposal facility should be built, and water quality parameters should be periodically measured and recorded. The sediment chromium and copper concentrations recorded in the study exceeded the TEL values reported by [3] , and the nickel concentrations exceeded both the TEL and PEC values.

Dam Lakes

Seyhan Dam Lake, built across the Seyhan River, has been in operation since 1956. It is an important dam lake in the Mediterranean Region, and was built for the purposes of flood control, irrigation and energy production. It also serves as an important recreation area for the city of Adana. [21] took sediment samples in each season of 2004-2005 (October, February, April and July) at five sampling stations in the dam lake. The sediments were seen to have metal concentrations in decreasing order as Ca > Fe > K > Mn > Na > Cr > Zn > Cu > Cd. The researchers reported that the probable reasons for the increased Cd and Cr concentrations seen in these sediment samples were agricultural activities and chrome mining in the area. Of the sediment metal concentrations observed in this study, the mean cadmium concentration was greater than the TEL value and the mean chromium concentration was greater than both the TEL and the PEC values [3] .

Porsuk and Enne Dam Lakes, located in Kütahya, are freshwater reservoirs that are highly polluted because of nearby industrialization and urbanization. There are also many boron deposits and thermal springs in the region surrounding these dam lakes. A study conducted by [22] examined heavy metal concentrations in sediment samples from both dam lakes and their bioaccumulation factors (BAFs) in various tissues (muscle, skin, gill, liver and intestine) of the common carp (Cyprinus carpio). The orders of sediment heavy metal concentrations were recorded as Fe > Mn > Zn > Ni > Pb > Cr > Cu > Cd > Se in Enne Dam Lake and Fe > Zn > Mn > Ni > Pb > Cr > Cu > Cd >

Se > Ag in Porsuk Dam Lake. There were no significant differences between sediment heavy metal concentrations in the two dam lakes. Hg concentration in the sediment of both reservoirs was lower than the detection limit. Sediment Cd and Pb concentrations were recorded respectively as 5.04 and 88.96 mg · kg^{-1} in Enne Dam Lake and 3.36 and 90.00 mg · kg^{-1} in Porsuk Dam Lake. The relatively high Cd sediment concentrations in both reservoirs could be a result of direct contamination of the water by Cd or the geochemical structure of the region. While the mean concentrations of zinc in Porsuk Dam Lake and cadmium in Enne Dam Lake greatly exceed both the TEL and PEC values, the mean concentrations of chromium, lead and nickel in both reservoirs, with the addition of cadmium in Porsuk Dam Lake, exceed only the TEL values reported by [3] .

Atatürk Dam Lake, on the Euphrates River, is the largest reservoir in Turkey. It is used for irrigation and electrical energy production purposes. It has a surface area of about 81,700 ha and a volume of 48,700,000,000 m^3, and 28 fish species are known to inhabit its waters. Industrial development and anthropogenic waste affect this reservoir, and the population in this region has recently increased [23] [24] . In a study conducted in 1997 by [23] , heavy metal concentrations (Cd, Co, Cu, Fe, Hg, Mn, Mo, Ni, Pb and Zn) were examined in the water, sediment and several fish species (Acanthobrama marmid, Chalcalburnus mossulensis, Chondrostoma regium, Carasobarbus luteus, Capoetta trutta and Cyprinus carpio) from Atatürk Dam Lake. Of the heavy metals tested for, Cd, Co, Hg, Mo and Pb were not detected in either water, sediment or fish samples, while Ni was observed to be at undetectable levels in the fish samples. Cu, Fe, Mn and Ni concentrations in the sediments from Akpinar station were higher than those from Bozyazi station. The researchers concluded that there was a general absence of serious heavy metal contamination in the reservoir, and the concentrations of elements that were observed were attributable to geological sources. Of the values recorded in the study, the nickel concentrations from Akpınar station exceeded both the TEL and PEC values, while those from Bozyazı station exceeded only the TEL for sediment levels reported by [3] .

A later study conducted by [24] observed trace element concentrations in water, sediment and fish samples (Carassius auratus, Capoeta trutta and Cyprinus carpio) collected from Atatürk Reservoir. The accumulation orders of trace elements were as follows: Fe > Zn

> Cu = Se in the water, and Fe > Mn > Ni >Zn > Cu > Se in the sediment. The accumulation order of trace elements in the tissues of the fish species was similar to one another. The researchers indicated that measures should be taken to preserve the present situation in the reservoir. The mean nickel concentration value obtained from the sediment samples were compared with SQG [3] , and it was found that, similar to the study undertaken in the same area by [23] , the value exceeded both TEC and PEC values.

Demirköprü Dam Lake, located in Manisa province on the Gediz River in western Anatolia, has a maximum capacity of 1,320,000,000 m^3 and an area of 47.66 km^2. Contamination studies must be performed on the reservoir at regular intervals as the reservoir is used for agricultural irrigation purposes and has a significant impact on both the environment and public health. In a 2005 study undertaken by [25] , heavy metal concentrations in the surface water, sediment and fish (Cyprinus carpio) were measured at two different stations on the reservoir. It was observed that heavy metal (Cd, Cr, Cu, Fe, Ni, Pb) concentrations in the water generally did not exceed six different guidelines. Heavy metal concentrations in the sediments decreased in sequence as follows: Fe > Ni > Cu > Cr > Pb > Cd. Only the cadmium concentrations in the sediment exceeded the TEL concentration values indicated by [3] .

Avsar Dam Lake, constructed in western Turkey to help ease water problems in the Gediz Basin, is actively fished by local inhabitants. The ecosystem's quality has been compromised as a result of agricultural and human activities. In a study conducted by [26] , concentrations of certain heavy metals (Cd, Cr, Cu, Fe, Ni and Pb) were seasonally observed in the water, the sediment and some tissues of Cyprinus carpio from the dam lake. The heavy metal concentrations in the sediment samples were as follows: Cd: 0.34 - 1.23 mg · L^{-1}, Cr: 9.41 - 19.9 mg · L^{-1}, Cu: 18.2 - 38.4 mg · L^{-1}, Fe: 19,680 - 28,560 mg · L^{-1}, Ni: 19.8 - 39.4 mg · L^{-1} and Pb: 0.64 - 6.35 mg · L^{-1}. Heavy metal concentrations in the lake sediments decreased in the order: Fe > Ni > Cu > Cr > Pb > Cd. The researchers maintained that the reservoir may be at risk in the future, depending on regional agricultural development. The researchers recommended that contamination studies be performed at regular intervals, as the reservoir is used for agricultural irrigation purposes and its condition affects both the environment and public health. The sediment metal concentrations were compared with the SQG [3] , and it was found that both the cadmium and nickel concentrations exceeded TEL values.

Geyik Dam Lake, which was constructed on the Sarıçay River in 1988 to provide irrigation and industrial water, supplies 38 hm³ of domestic water to the town of Milas. In a study conducted on the reservoir, new information was obtained on the accumulation of heavy metals (Cd, Co, Cu, Fe, Mn, Ni, Pb, and Zn) in fish tissue (Cyprinus carpio and Carassius carassius), sediment, and water from the reservoir. Sediment samples from the study showed high Fe concentrations in winter and summer, but low concentrations of Cd and Pb in summer, and likewise, low Co and Cu in winter [27] . The mean concentration of nickel in the sediment was much higher than both TEL and PEC values, and the highest value for copper concentration exceeds the TEL value reported by [3] .

Çatören Dam Lake is in Seydisuyu Basin in the Kırka district of Eskişehir province. This is one of Turkey's most important boron mining regions. In the reservoir some limnological parameters as well as macro-micro accumulations of certain elements in the water, the sediment, and some tissues of Cyprinus carpio, Cyprinus carpio var. specularis and Tinca tinca, which are important fish for regional trade [28] . The study noted that total P, Zn, Mn, As and B sediment concentrations detected at the output of the reservoir were significantly higher than those detected at the input. The highest As and Zn values were recorded at 12.4 mg · kg⁻¹ and 32 mg · kg⁻¹, respectively. The cadmium concentration in the sediment exceeded both the TEL and PEC values set by [3] .

Rivers and Streams

Some heavy metal concentrations in different rivers and sediment samples (Turkey) are presented in Table 3.

The Gediz and Buyuk Menderes (BM) Rivers, which have great economic importance for Turkey, in order to determine the extent of environmental contamination [29] . The study reported significant contamination levels in both rivers, with high levels of Pb, Cr, Mn and Zn being reported in the Gediz River and of Co, Mn, and Zn in the BM River. This contamination was likely caused by industrial waste, according to speciation studies. While the chromium and nickel concentrations in both rivers exceeded both TEL and PEC values, the copper and lead concentrations in both the rivers exceeded only the TEL and the mean zinc value in the Gediz River exceeded only the TEL value as reported by [3] .

In a separate study undertaken on the Sakarya River, [30] collected water and sediment samples from three different stations along the river between May and September 2003. The results showed differences between different seasons, sampling stations, and nature of the samples (sediment or water). The mean levels of heavy metals in the sediment samples (copper, nickel, chromium, lead, cadmium, zinc) are as follows: 4.630 $\mu g \cdot g^{-1}$, 13.520 $\mu g \cdot g^{-1}$, 8.780 $\mu g \cdot g^{-1}$, 2.640 $\mu g \cdot g^{-1}$, nd, 9.990 $\mu g \cdot g^{-1}$, and in the water samples as: 0.85 $\mu g \cdot g^{-1}$, 1.050 $\mu g \cdot g^{-1}$, 0.027 $\mu g \cdot g^{-1}$, 1.786 $\mu g \cdot g^{-1}$, 0.236 $\mu g \cdot g^{-1}$ and 0.173 $\mu g \cdot g^{-1}$, respectively. The researchers concluded that, while several studies have shown that the Sakarya River has a high level of contamination, when the total heavy metal concentrations in sediments are taken into account, the lower part of the river is seen to have a higher level of contamination than the upper part. Of the heavy metal concentrations in the sediment, the mean concentration of nickel is much higher than the TEL reported by [3] .

The Yesilirmak River has a total basin area of 2352.8 km², and the river is 519 km long. A large part of the river flows through the city of Tokat. Concentrations of trace metals in sediment samples from the river were as follows: Cu 37.9 mg $\cdot$ kg^{-1}, Mn 392.2 mg $\cdot$ kg^{-1}, Zn 126.2 mg $\cdot$ kg^{-1}, Fe 3726 mg $\cdot$ kg^{-1}, and Pb 29.6 mg $\cdot$ kg^{-1} [31] . While among the total concentrations of trace metals analyzed in the study, copper and zinc were shown to exceed the TEL values given by [3] , in another study of the same area, [32] found that maximum metal concentrations for copper (38.7 $\mu g \cdot g^{-1}$) exceeded [3] 's reported TEL value for copper, and mean nickel values (79.2 $\mu g \cdot g^{-1}$) exceeded both the TEL and PEC values for nickel.

The Tigris River has its start in the mountains of Eastern Anatolia and flows southeastwards from Turkey into Iraq. Samples taken by [33] from study sites on the river and a reference site on Resan Creek between June 2000 and May 2001 returned metal levels of the sediments sequenced as: Fe > Mn > Cu > Co > Zn. The highest copper and nickel concentrations in the sediment equaled [3] 's recommended TEL highest values, and exceeded the PEC values.

The Dipsiz Stream is 88 km long and yields about 1000 L/s⁻¹. It flows into the Buyuk Menderes River in southwestern Turkey. In three surveys carried out on the stream in April, July and October 2004 [34] , heavy metal concentrations (Cd, Cr, Cu, Pb and Zn) were measured in the water, sediment and muscle and gill tissues of Leuciscus cephalus in Yatagan Basin, the site of a geothermal power plant. The mean sediment concentrations of cadmium and lead were shown to be much higher than the TEL reported by [3] .

Table 3: Heavy metal concentrations (mg · g^{-1} DW) in sediment samples from different rivers and streams in Turkey

Heavy metal cons. Rivers and Streams	As	Cd	Cr	Cu	Pb	Hg	Ni	Zn	References
Büyük Menderes River	-	-	165.0 ± 7.01,2	137.0 ± 5.01	54.0 ± 8.01	-	315.0 ± 25.01,2	120.0 ± 10.0	[29]
Gediz River	-	-	200.0 ± 6.01,2	140.0 ± 3.01	128.0 ± 15.01	-	106.0 ± 10.01,2	160.0 ± 15.01	
Kızılırmak River	-	0.92 ± 0.031	-	9.95 ± 2.65	0.62 ± 0.04	-	47.91 ± 6.431	34.62 ± 6.46	[35]
Sakarya River (in the lower part)	-	nd	9.47 ± 2.7	5.63 ± 1.70	4.31 ± 0.40	-	18.17 ± 2.101	10.55 ± 2.20	[30]
TigrisRiver	-	nd	-	25.39 - 194.531,2(H)	nd	-	36.68 - 170.831,2(H)	17.09 - 66.26	[33]
Ye ilırmak River		0.55		38.71	17.3		79.21,2	45.5	[32]
				37.91	29.6			126.21	[31]
Dipsiz Stream	-	0.80 ± 0.601	19.70 ± 15.60	13.00 ± 9.00	83.60 ± 56.201	-	-	37.00 ± 26.00	[34]
Delice Stream (Kızılırmak Basin)	-	-	74.4 - 237.41,2	7.11 - 57.31(H)	27.99 - 39.71(H)	4.16 - 14.811,2	-	86.2 - 170.61(H)	[36]

Felent Stream	As	Cd	Cr	Cu	Pb	Hg	Ni	Zn	Reference
Upstream	15.56 - 119.90[1,2](H)	-	48.38 - 62.69[1]	9.62 - 19.53	11.73 - 230.75[1,2](H)	-	-	39.31 - 343.25[1](H)	[28]
Lentic	27.12[1]	-	53.47[1]	22.29	11.29	-	-	54.69	
Downstream	14.84 - 18.53[1]	-	33.42 - 46.29[1](H)	15.47 - 163.48[1,2](H)	8.52.123.66[1](H)	-	-	56.13 - 138.00[1](H)	
SQGs	As	Cd	Cr	Cu	Pb	Hg	Ni	Zn	Reference
TEL1	5.9	0.596	37.3	35.7	35.0	0.174	18.0	123	[3]
PEC2	33.0	4.98	111.0	149.0	128.0	1.06	48.6	459.0	

[1]Every superscript with 1 for heavy metal concentrations mean the concentration exceeds the TEL value; [2]Every superscript with 2 for heavy metal concentrations mean the concentration exceeds the PEC value.

The Kızılırmak River the longest river within the borders of Turkey (1355 km) is currently used as a source of drinking water for Ankara, the capital of Turkey. The study carried out in Kızılırmak River (in April and October 2005) has been conducted to determine cadmium, cobalt, copper, chromium, iron, manganese, nickel, lead and zinc concentrations in the five different tissues of carp (gills, muscle, liver, kidney and gonads), and in the surface water, sediment samples, and aquatic plant from the study area. The highest sediment heavy metal concentration was recorded for Fe, and the lowest for Cd and Pb. The origin of the elements could be attributed to the geological environment surrounding the river, and some industrial facilities near the river. Results of correlation coefficient analysis gained from the study suggested that metal concentrations in the sediment and aquatic plant are the most curucial factors governing the metal body concentration of fish [35] . The cadmium and nickel concentrations in the sediment exceeded the TEL value for these metals outlined by [3] . The accumulation of heavy metals such as Pb, Hg, Co, Cr, Cu, Zn, and Br in Delice Stream, which is a tributary of the Kızılırmak River, were studied in water and sediment samples as well as samples of muscle and gill of three fish species (Leuciscus cephalus, Capoeta tinca, Capoeta capoeta). The samples were taken from four different stations in February, May and August 2008. The metal concentrations in the sediment samples were sequenced as: Cr > Zn > Pb > Cu > Co > Hg > Br [36] . According to the researchers, if these water bodies are to be used as drinking water and agricultural water sources, they must be biomonitored in order to keep tabs on their heavy metal concentrations. The chromium and mercury concentrations in the sediment exceeded both the TEL and the PEC values for these metals outlined by [3] . The lead, copper and zinc concentrations (high values) exceeded only the TEL.

Felent Stream is one of the most important branches of Porsuk Stream in the Sakarya River Basin. It is used for irrigation, industrial water supply, domestic waste disposal and fishing [28] . Seasonal sediment samples were taken from Felent Stream in 2011 at three different locations: the mining-agricultural section (upstream), the lenthic section (a reservoir) and the urban section (downstream). The metal concentrations in these sediment samples were then compared with SQG [3] . The comparison showed that the upstream and downstream section's heavy metal concentration values were similar to those indicated by [3] . As for the lenthic section, the arsenic and chromium concentrations were seen to be higher than the TEL values given by [3] .

According to water quality regulations [37] , Turkey's inland waters are divided into four classes. While some of the lakes mentioned in this review, such as Beyşehir, Uluabat and Yeniçağa Lakes, have heavy metal concentrations higher than permissible for drinking water, others such as Kovada and Işıklı Lakes don't have this problem. Additionally, the sediment heavy metal levels in Uluabat and Yeniçağa Lakes were found to be extremely high, and in this type of lake, create a risk for food contamination, as in the case of the crayfish (Astacus leptodactylus) that are caught in Yeniçağa Lake.

Although the geochemical structure of the region does play a role in the increase of heavy metal levels in lake sediment, the main cause has been shown to be terrestrial inputs originating from anthropogenic (domestic and industrial) sources which are delivered via rivers and rainfall. Especially harmful are effluents from urban settlements on the lake's shores and agricultural development in the drainage basin. Studies done in different years on the same aquatic systems have shown that heavy metal concentrations in the water and sediment have been on the increase. This result points to the fact that contamination is continuing and the measures that have been taken are not effective enough. When the heavy metal concentrations in the lake sediment mentioned in this review are examined in light of the TEL and PEC levels for sediment reported by [3] , it can be seen that in many cases, the nickel concentrations exceeded the TEL values, followed by cadmium. It is common knowledge that ore, smelter and refinery waste are the major causes of nickel contamination. Nickel, which is widely used in the electronics, steel, battery and food production industries and can be found in dissolvable form in aquatic environments, bonds to clay minerals or organic particles (especially those such as humic and fulvic acid). Cadmium is used in the paint and battery industries, in the coating-galvanizing and aircraft industries, due to its non-cor- rosive properties, in insecticide formulation, and as a stabilizer in plastics production. Because it is the element with the highest potential to dissolve in water, its diffusion rate in natural environments is extremely high [1] . The results of this review indicate that heavy metal contamination in lake systems in Turkey should be taken into account when developing management strategies for protecting aquatic environments.

The contamination sources that lead to high heavy metal concentrations in wetland and reservoir sediment are no different from those that affect other aquatic systems. For many reservoirs, for example Demirköprü, Avşar, Seyhan and Geyik reservoirs, it has been suggested that the heavy metal levels in the sediment result from agricultural and domestic waste discharges. For others, primarily Atatürk Reservoir, geological sources are thought to play a role as the main contaminating element. For example, for Seyhan Reservoir, chrome mining, and for Çatören Reservoir, boron processing were found to contribute to the increase of some sediment heavy metal concentrations. In highly polluted reservoirs like Porsuk and Enne Reservoirs, on the other hand, industrialization and urbanization activities are the primary culprit. Among the reservoirs discussed (Demirköprü and Avşar Reservoirs) concentrations of especially Cd, Cr, Ni and Pb in the fish that are caught for food exceed tolerable values delineated by international institutions. When the heavy metal concentrations in the sediment of the wetlands and reservoirs investigated in this review are examined according to the threshold effect level (TEL) and probable effect concentrations (PEC) for sediment levels reported by [3] , it is clear that in the majority of cases (Yedigöller, Porsuk and Enne Reservoirs, Avşar Reservoir, Atatürk Reservoir and Geyik Reservoir), just as in the lake ecosystems examined, of the heavy metals, nickel concentrations exceeded the TEL value, followed by cadmium. In this context, for reservoir management purposes it is necessary to determine the condition of the reservoirs and initiate monitoring programs.

When examining heavy metal levels in the sediment of rivers and streams, it has been noted that limited studies have been done on this topic. Of the areas that are included in such studies, it has been reported that especially for the Büyük Menderes, Gediz and Yeşilırmak river basins, industrial plants, agricultural lands and sites are responsible for heavy metal contamination. When the sediment heavy metal concentrations for the rivers and streams discussed in this study were evaluated in light of the TEL and PEC for sediment levels indicated by [3] , it was determined that nickel, as well as other heavy metal concentrations like chromium, copper, mercury and lead, exceeded TEL values. In rivers like the Büyük Menderes and Gediz, which are among those with basins especially at risk for heavy contamination, chromium and nickel concentrations exceeded both TEL and PEC values. Thus, in the context of different analyses of heavy metals in the sediment of rivers, it is seen that the most effective measures to improve the quality of the river basin are to control of the quantity and concentration of input contaminants and reduce the stream's pollution load.

In conclusion, there has been considerable attention given to sediment heavy metal contamination in Turkey's freshwater ecosystems over the past 15 years. Terrestrially derived wastewater discharge, agricultural and industrial run-off and regional geological characteristics are known to be the most common sources of heavy metals in the sediments of these freshwater ecosystems. All five types of freshwater aquatic systems in Turkey are contaminated with heavy metals according to the TEL and PEC values. Moreover, the tissues of aquatic biota in the study areas have been reported as having high levels of heavy metals in them. For this reason, site-specific information should be collected to supplement the use of SQGs for evaluating sediment quality on a regional and national basis.

RESEARCH REGARDING NUTRIENT RELEASE FROM SEDIMENT IN INLAND AQUATIC ECOSYSTEMS

In aquatic environments, phosphorus transfer occurs between water and sediment as a result of several physical, chemical and biological events. Transfer of phosphorus from the sediment to the water is known as internal phosphorus loading. Sediment is very important for the nutrient dynamics of shallow lakes. Even in lakes where the external loading has been reduced, internal phosphorus loading from the sediment has in some cases been seen to prevent improvements in lake water quality [38] [39] .

Phosphorus release amounts are estimated by comparing total filterable orthophosphate concentrations to the sediment porewater as well as overlying waters, with the water content of the sediment and the diffusion coefficient depending on the temperature. Physical, chemical and biological measures must be taken in aquatic systems in order to control eutrophication by reducing internal phosphorus load. Phosphorus release from the sediment to the lake water through molecular diffusion can generally be estimated using Fick's First Law. Accordingly, the formula specified by [40] is widely used for this purpose. Phosphorus release is dependent on a number of factors. Considered particularly important are redox sensitive mobilization from the anoxic zone a few millimeters or centimeters below the sediment surface and microbial processes. However, the mechanisms contributing to phosphorus release are sometimes specific to an individual lake [38] [39] [41] .

The number of studies conducted in the last 15 years regarding the release of nutrients from the sediment to the water of some lakes and ponds in Turkey is extremely low. The first study of this type was done at Kırkgız Pond (West Pond), one of the five Sakaryabaşı Karst Springs. These Karst Springs in Central Anatolia are part of a confined/semi-confined karst aquifer, which also has a thermal component. Kırkgız Pond was naturally converted to a pond due to the presence of macrophytes in the spring. The pond supplies water to the Rainbow Trout Culture and Research Station, which produces approximately 40 tons of fish per year. The above-men- tioned study on Kırkgız Pond was conducted by [42] in 2000 and 2001 in order to establish changes in concentrations of phosphorus and iron in the pondwater and porewater in the littoral zone of the pond. Soluble reactive phosphorus concentrations were four to five times higher in the porewater than the pondwater, making diffusion of SRP from porewater to pondwater a distinct possibility. According to the results of the study, it seems that it would be easier to control phosphorus concentration in spring and autumn than in the warm summer months. The aim of a different study conducted at West Pond, which also supplies water to the Sakaryabaşı Fish Culture and Research Station, was to quantitatively estimate the phosphorus release from the sediment due to diffusion. The highest release rate was recorded in October 2000 as 22 $\mu g \cdot m^{-2} \cdot d^{-1}$, and the lowest was 3 $\mu g \cdot m^{-2} \cdot d^{-1}$ in January 2001. Thus, it was quantitatively shown that the sediment doesn't have a significant effect on the nutrient levels of West Pond. It was concluded that the pond's high calcium content and the sediment's limy composition were the important elements causing phosphorus release from the sediment to remain at low levels, depending on the water temperature and the pH values [43] . A third study, also undertaken on West Pond, had multiple aims: to determine phosphorus release from the sediment to the water using vertical sections (5.0 cm sampling intervals across 20 cm sediment depth), to observe seasonal changes (positive-upwards and/or negative-downwards) in the phosphorus fractional composition of the eutrophicated West Pond sediment with phosphorus cycling between the sediment and the water, and to determine which fraction or fractions potentially have the greatest effect on phosphorus release and the pond's eutrophication process. The mean release values of phosphorus from the sediment in West Pond, measured at similar depths and in months representing the seasons (April, July, October, January), varied between 9.19 (January, 0 - 5 cm) and 119.08 $\mu g \cdot m^{-2} \cdot d^{-1}$ (October, 0 - 5 cm). The mean negative phosphorus release values according to months and depths were estimated as 1.25 $\mu g \cdot m^{-2} \cdot d^{-1}$ (July, 5 - 10 cm) and 46.45 $\mu g \cdot m^{-2} \cdot d^{-1}$ (October, 10 - 15 cm). Phosphorus fractions were observed to be distributed in the pond sediment as follows: total organically bound phosphorus fraction (Org $\approx$ P) > calcium bound phosphorus fraction (Ca $\approx$ P) > carbonate bound phosphorus fraction (CO $\approx$ P) > iron + aluminum bound phosphorus fraction (Fe + Al $\approx$ P). Similar to previous research results, the negative phosphorus release dependent on months and depths has quantitatively shown a very low flux [44] . Another study of West Pond on the topic of nitrogen release from the sediment to the water, estimated inorganic nitrogen (ammonium-nitrate) flux in months representing the seasons (April, July, October, January. Negative ammonium and nitrate flux (retention) values were observed to change between 809.48 and 2069.52 $\mu g \cdot m^{-2} \cdot day^{-1}$, at 2053.77 to 7718.10 $\mu g \cdot m^{-2} \cdot day^{-1}$, respectively. The minimum inorganic nitrogen (ammonium-nitrate) flux value was estimated in July, and the maximum in January. The study quantitatively suggests that West Pond's sediment doesn't function as a source of nitrogen, but rather, a trap. This conclusion points to the importance of the management of macrophytes, which are found in large concentrations in the littoral zone and function as a mechanism for nitrogen removal, in pond rehabilitation efforts [45] . It can be concluded from examining the scientific data on the sediment of West Pond that the sediment is a sink, not a source, of phosphorus and inorganic nitrogen. Thus, the best management technique for the pond is conservation of the aquatic macrophytes' function in preventing and/or suppressing the dissolved inorganic phosphorus and nitrogen release from the sediment to the overlying water.

Mogan Lake, an important recreational area for metropolitan Ankara, is designated with environmental protection status as "Gölbaşı

Specially Protected Area." The chief source of phosphorus in the region is waste-water, which contains detergents and fertilizers used in the nearby agricultural areas. A study was undertaken between July 2004 and June 2005 to examine the seasonal and spatial patterns of Mogan Lake's littoral sediment phosphorus and the potential for its release into the water. Sediment phosphorus release fluctuated between 0.002 and 0.062 $\mu g \cdot m^{-2} \cdot d^{-1}$ within the research period, and reached minimum and maximum values in November and June, respectively. By examining the estimated values of phosphorus release into the lake, the study has quantitatively shown that the sediment had only a minimal effect of on the trophic level of the lake. Organically-bound phosphorus fractions were estimated to make up the largest proportion of fractions preventing phosphorus release into the lake (Org $\approx$ P: 36%), followed by calcium-bound (Ca $\approx$ P: 35%), iron + aluminium-bound (Fe + Al $\approx$ P: 16%) and carbonate-bound ($CO_3 \approx$ P: 13%) phosphorus fractions. Internal phosphorus load was estimated to be very low, and thus, it is suggested that sediment dredging of the lake would not affect its trophic level, and would only deepen the lake [46] . A concurrent study conducted in Mogan Lake between September 2005 and August 2006 [47] examined vertical distributions of total phosphorus (TP), phosphorus fractions, iron and organic matter in the littoral sediment in a macrophyte-dominated, clear water state. Benthic macroinvertebrates and total bacteria in the sediment also were determined in the study. It was concluded that no clear seasonal or depth-related (0 - 20 cm) patterns could be seen in sediment concentrations for the measured parameters. Phosphorus release was quantitatively quite low, with negative phosphorus release ($-0.132 \mu g \cdot m^{-2} \cdot day^{-1}$) recorded during the summer months. Sediment TP concentrations ranged between 675.00 and 1463.80 $\mu g \cdot g^{-1}$ dry weight (DW), and the trophic level of the lake was determined as eutrophic. Here, inorganic phosphorus fractions made up the largest fraction (63%), with organic-bound phosphorus (Org $\approx$ P) constituting 37% of total phosphorus. The low occurrence of benthic macroinvertebrates (510 - 850 individuals/ m^{-2}), which are dependent on sediments with high iron content and low levels of organic matter (5.42% - 13.30%), affected sediment phosphorus retention. Bacterial presence in the surface sediment appeared to be positively correlated with temperature; however, anoxic conditions weren't present in the overlying water, pointing to the possibility that the bacteria retained phosphorus in their cell structures.

A third study undertaken in macrophyte-dominated, clearwater state eutrophic Mogan Lake, which has extremely low positive and negative diffusional phosphorus release, examined variations in total dissolved phosphorus (TDP), soluble reactive phosphorus (SRP) and total iron (TFe) concentrations and pH values in the littoral sediment porewater between 0 - 20 cm depth at 5 cm intervals over a period of eleven months. The objective of this study was to document whether these values differ according to season and depth. The findings are discussed in terms of the dissolved oxygen, pH and redox potential of the overlying water and low-release data. The continued aerobic-environment low phosphorus release from Lake Mogan's sediment is due in part to low SRP concentration gradients between the overlying water and the porewater. Keeping this in mind, it should be noted that changes to the in-lake phosphorus cycling mechanisms could potentially be implemented through monitoring SRP variations in the porewater and the overlying water, and lake sediments would thus become net sources of SRP rather than net sinks. This concept is a potentially important one for lake sediment management strategies [48] .

From these three studies involving sediment phosphorus release in Mogan Lake, it can be concluded that the sediments of this lake must be monitored over time, taking management measures and continued external phosphorus loading into account, to improve its management and promote the continuation of a clear water state. Consequently, two steps should be taken in order to control the eutrophication process in Lake Mogan: first, the sediment should be examined to determine its adsorption capacity, and second, a program should be implemented which monitors the lake's trophic level by recording variations in the lake water's total filterable orthophosphate concentrations.

Küçük Çekmece Lagoon, located in the Marmara Region 15 km to the west of Turkey's largest metropolis, Istanbul, is a shallow sea-level wetland water source used for various purposes. It has gained widespread international importance due to its biological diversity. In a study comparing sediment nutrient release in the lagoon under field and laboratory conditions, sediment samples were taken monthly between October 2006 and February 2008. The highest readings for nitrate (NO_3-N) and orthophosphate (o-PO_4) obtained under field conditions were 3.39 mg/L^{-1} and 6.62 mg/L^{-1}, respectively. However, under laboratory conditions, the highest value of NO_3-N was measured as 3.41 mg/L^{-1}, while the highest value of o-PO_4 was 2.95 mg · L^{-1}.

Using these results, the highest values for nitrate and orthophosphate flux were calculated as 3035 and 1487 mg $\cdot$ m^{-2} $\cdot$ day^{-1}, respectively. Under field conditions, significant variation was seen in the measured nutrient values between the deepest and costal stations. Under laboratory conditions, however, the measured values did not generally show such a difference. The data from the study point toward a nutrient load in the lagoon characteristic of highly eutrophic lakes. According to the study, sediment fluxes account for 45% of the nitrogen load and 85% of the phosphorus load entering the Küçük Çekmece Lagoon annually. This result shows that sediment flux plays an important role in the lagoon's eutrophication process [49] . It is a scientific fact that the pollution load of Küçük Çekmece Lagoon is continually increasing as a result of the discharge of untreated industrial and domestic waste water directly into the lagoon or the streams that feed it. With this study of the lagoon, it has been suggested that pollutants accumulate not only in the water, but also in the sediment structure, having a negative effect on the sustainability of the lagoon.

Sediment has been shown to have a significant effect on nutrient transfer in lakes, especially shallow ones. When phosphorus enters the sediment of a lake, it is employed in the chemical and biological processes of the sediment. Later, it either becomes permanently deposited in the sediment or is released by various processes and returns to the water column in dissolved form through the porewater. It should be emphazised, however, that sediment can differ drastically from one lake to another, and even within the same lake, may have highly variable chemical composition. Factors such as dry weight, organic content, and the presence of iron, aluminum, manganese, calcium, clay and other elements with the capacity to bind and release phosphorus may affect interactions between the water and the sediment. It can be understood from this review that there have been an extremely limited number of studies on this topic conducted in Turkey. Due to their significant effect on lake water concentrations, awareness of sediment-water interaction and the processes behind phosphorus retention and release in the sediment is crucial in order to understand shallow lakes' functioning and determine effective management strategies.

REFERENCES

1. Topçuoğlu, S. (2005) Denizel Biyota Örneklerinde Ağır Metal Kontaminasyonu. Deniz Kirliliği. In: Güven, K.C. and Öztürk, B., Eds., TÜDAV Yayınları, İstanbul, 205-225.

2. Boyd, C.E. and Tucker, C.S. (1998) Pond Aquaculture Water Quality Management. Kluwer Academic Publishers, 698 p. http://dx.doi.org/10.1007/978-1-4615-5407-3

3. MacDonald, D.D., Ingersoll, C.G. and Berger, T.A. (2000) Development and Evaluation of Consensus-Based Sediment Quality Guidelines for Freshwater Ecosystems. Archives of Environmental Contamination and Toxicology, 39, 20-31.http://dx.doi.org/10.1007/s002440010075

4. Altındag, A. and Yigit, S. (2005) Assessment of Heavy Metal Concentrations in the Food Web of Lake Beysehir, Turkey. Chemosphere, 60, 552-556.

5. Kır, I., Ozan Tekin, S. and Tuncay, Y. (2007) Kovada Gölü'nünsu vesedimentindekibazı ağırmetallerin mevsimsel de- ğişimi. EU Journal of Fisheries & Aquatic Sciences, 24, 155-158.

6. Mendil, D. and Uluözlü, D.Ö. (2007) Determination of Tracemetal Levels in Sediment and Five Fish Species from Lakes in Tokat, Turkey. Food Chemistry, 101, 739-745.http://dx.doi.org/10.1016/j.foodchem.2006.01.050

7. Özmen, H., Külahci, F., Cukurovali, A. and Dogru, M. (2004) Concentrations of Heavy Metal and Radio Activity in Surface Water and Sediment of Hazar Lake (Elazig, Turkey). Chemosphere, 55, 401-408. http://dx.doi.org/10.1016/j.chemosphere.2003.11.003

8. Karadede-Akın, H. (2009) Seasonal Variations of Heavy Metals in Water, Sediments, Pondweed (P. Pectinatus L.) and Freshwaterfish (C. c. umbla) of Lake Hazar. Elazığ-Turkey. Fresenius Environmental Bulletin, 18, 511-518.

9. Kurun, A., Balkıs, N., Erkan, M., Balkıs, H., Aksu, A. and Erşan, M.S. (2010) Total Metal Levels in Crayfish Astacus leptodactylus (Eschscholtz, 1823), and Surface Sedimentsin Lake Terkos, Turkey. Environmental Monitoring and Assess- ment, 169, 385-395.http://dx.doi.org/10.1007/s10661-009-1181-5

10. Dalkıran, N., Karacaoğlu, D., Dere, Ş., Sentürk, E. and Torunoğlu, T. (2006) Factors Affecting the Current Status of a Eutrophic Shallow Lake (Lake Uluabat, Turkey): Relationships between Water Physical and Chemical Variables. Che- mical Ecology, 22, 279-298. http://dx.doi.org/10.1080/02757540600856229

11. Anonymous (2013) http://www.turkiyesulakalanlari.com/sulak-alanlar

12. Barlas, N., Akbulut, N. and Aydoğan, M. (2005) Assessment of Heavy Metal Residues in the Sediment and Water Samples of Uluabat Lake, Turkey. Bulletin of Environmental Contamination and Toxicology, 74, 286-293. http://dx.doi.org/10.1007/s00128-004-0582-y

13. Elmacı, A., Teksoy, A., Topaç, O.F., Özengin, N., Kurtoğlu, S. and Başkaya, H.S. (2007) Assessment of Heavy Metals in Lake Uluabat, Turkey. African Journal of Biotechnology, 6, 2236-2244.

14. Arslan, N., Koç, B. and Çiçek, A. (2010) Metal Contents in Water Sediment and Oligochaeta-Chironomidae of Lake Uluabat a Ramsar Site of Turkey. The Scientific World Journal, 10, 1269-1281. http://dx.doi.org/10.1100/tsw.2010.117

15. Tekin-Özan, S. and Aktan, N. (2012) Relationship of Heavy Metals in Water, Sediment and Tissues with Total Length, Weight and Seasons of Cyprinus carpio L., 1758 from Işikli Lake (Turkey). Pakistan Journal of Zoology, 44, 1405- 1416.

16. Tunca, E., Atasagun, S. and Saygı, Y. (2012) Yeniçağa Gölü'nde (Bolu-TÜRKİYE) su, sediment ve kerevitteki (Astacus leptodactylus) bazı ağır metallerin birikimi üzerine bir ön çalışma. Ekoloji, 21, 68-76. http://dx.doi.org/10.5053/ekoloji.2012.838

17. Benzer, S., Arslan, H., Uzel, N., Gül, A. and Yılmaz, M. (2013) Concentrations of Metals in Water, Sediment and Tissues of Cyprinus carpio L., 1758 from Mogan Lake (Turkey). Iranian Journal of Fisheries Sciences, 2, 45-55.

18. Başyiğit, B. and Tekin-Özan, S. (2013) Concentrations of Some Heavy Metals in Water, Sediment and Tissues of Pikeperch (Sander lucioperca) from Karataş Lake Related to Physico-Chemical Parameters, Fish Size and Seasons. Polish Journal of Environmental Studies, 22, 633-644.

19. Dural, M., Goksu, M.Z.L. and Ozak, A.A. (2007) Investigation of Heavy Metal Levels in Economically Important Fish Species Captured from the Tuzla Lagoon. Food Chemistry, 102, 415-421. http://dx.doi.org/10.1016/j.foodchem.2006.03.001

20. Arslan, N., Tokatlı, C., Çiçek, A. and Köse, E. (2011) Determination of Some Metal Concentrations in Water and Sediment Samples in Yedigöller Region (Kütahya). Review of Hydrobiology, 4, 17-28.

21. Çevik, F., Göksu, L., Derici, B. and Fındık, Ö. (2009) An Assessment of Metal Pollution in Surface Sediments of Seyhan Dam by Using Enrichment Factor, Geoaccumulation Index and Statistical Analyses. Environmental Monitoring and Assessment, 152, 309-317.http://dx.doi.org/10.1007/s10661-008-0317-3

22. Uysal, K., Özden, Y., Çiçek, A. and Köse, E. (2010) Bioaccumulation Ratios of Sediment-Bound Heavy Metals of Porsuk and Enne Dam Lakes (Kütahya/Turkey) to Different Tissues of Common Carp (Cyprinus carpio). İstanbul University Journal of Fisheries and Aquatic Sciences, 25, 1-10.

23. Karadede, H. and Ünlü, E. (2000) Concentrations of Some Heavy Metals in Water, Sediment and Fish Species from the Atatürk Dam Lake (Euphrates), Turkey. Chemosphere, 41, 1371-1376. http://dx.doi.org/10.1016/S0045-6535(99)00563-9

24. Ural, M., Uysal, K., Çiçek, A., Köse, E., Koçer, M.A.T., Arca, S., Örnekçi, G.N., Demirol, F. and Yüce, S. (2011) De- termination of Trace Element Concentraions in Water, Sediment and Fish Species from the Atatürk Dam Lake (Eu- phrates), Turkey. Fresenius Environmental Bulletin, 20, 2036-2040.

25. Öztürk, M., Özözen, G., Minareci, O. and Minareci, E. (2008) Determination of Heavy Metals in of Fishes, Water and Sediment from the Demirköprü Dam Lake (Turkey). Journal of Applied Biological Sciences, 2, 99-104.

26. Öztürk, M., Özözen, G., Minareci, O. and Minareci, E. (2009) Determination of Heavy Metals in of Fish, Water and Sediment of Avşar Dam Lake inTurkey. Iranian Journal of Environmental Health Science Engineering, 6, 73-80.

27. Özdemir, N., Yılmaz, F., Levent Tuna, A. and Demirak, A. (2010) Heavy Metal Concentrations in Fish (Cypri- nuscarpio and Carassiuscarassius) Sediment and Water Found in the Geyik Dam Lake, Turkey. Fresenius Environ- mental Bulletin, 5, 798-804.

28. Çiçek, A., Tokatlı, C., Emiroğlu, Ö., Köse, E., Başkurt, S. and Sülün, Ş. (2013) Macro and Micro Element Con- centrations in Water, Sediment and Commercial Fishes of Çatören Dam (Eskişehir). Journal of Researh in Ecology, 2, 91-99.

29. Akcay, H., Oguz, A. and Karapire, C. (2003) Study of Heavy Metal Pollution and Speciation in Buyak Menderes and Gediz River Sediments. Water Research, 37, 813-822.http://dx.doi.org/10.1016/S0043-1354(02)00392-5

30. Dundar, M.S. and Altundag, H. (2007) Investigation of Heavy Metal Contaminations in the Lower Sakarya River Water and Sediments. Environmental Monitoring and Assessment, 128, 177-181. http://dx.doi.org/10.1007/s10661-006-9303-9

31. Tüzen, M. (2003) Determination of Trace Metals in the River Yeşilırmak Sediments in Tokat, Turkey Using Sequential Extraction Procedure. Microchemical Journal, 74, 105-110. http://dx.doi.org/10.1016/S0026-265X(02)00174-1

32. Mendil, D., Ünal, Ö.F., Tüzen, M. and Soylak, M. (2010) Determination of Trace Metals in Different Fish Species and Sediments from the River Yeşilırmak in Tokat Turkey. Food and Chemical Toxicology, 48, 1383-1392. http://dx.doi.org/10.1016/j.fct.2010.03.006

33. Karadede-Akın, H. and Ünlü, E. (2007) Heavy Metal Concentrations in Water, Sediment, Fish and Some Benthic Organisms from Tigris River, Turkey. Environmental Monitoring and Assessment, 131, 323-337. http://dx.doi.org/10.1007/s10661-006-9478-0

34. Demirak, A., Yilmaz, F., Levent Tuna, A. and Özdemir, N. (2006) Heavy Metals in Water, Sediment and Tissues of Leuciscus cephalus from a Stream in Southwestern Turkey. Chemosphere, 63, 1451-1458. http://dx.doi.org/10.1016/j.chemosphere.2005.09.033

35. Yılmaz, F. (2006) Bioaccumulation of Heavy Metals in Water, Sediment, Aquaticplants and Tissues of from Kızılırmak, Turkey. Fresenius Environmental Bulletin, 5, 360-369.

36. Akbulut, A. and Akbulut, N. (2010) The Study of Heavy Metal Pollution and Accumulation in Water, Sediment, and Fish Tissue in Kızılırmak River Basin in Turkey. Environmental Monitoring and Assessment, 167, 521-526. http://dx.doi.org/10.1007/s10661-009-1069-4

37. Anonymous (2004) http://www.cygm.gov.tr/CYGM/homepage/ regulations

38. Sondergaard, M., Jensen, J.P. and Jeppesen, E. (2001) Retention and Internal Loading of Phosphorus in Shallow, Eutrophic Lakes. The Scientific World, 1, 427-442.http://dx.doi.org/10.1100/ tsw.2001.72

39. Eckert, W., Didenko, J., Uri, E. and Eldar, D. (2003) Spatial and Temporal Variability of Particulate Phosphorus Fractions in Seston and Sediments of Lake Kinneret under Changing Loading Scenario. Hydrobiologia, 494, 223-229.http://dx.doi. org/10.1023/A:1025474517703

40. Shaw, J.F.H. and Prepas, E.E. (1990) Relationships between Phosphorus in Shallow Sediments and in the Trophogenic Zone of Seven Alberta Lakes. Water Resources, 24, 551-556.

41. Cooke, G.D.,Welch, E.B., Peterson, S. and Newroth, P. (1993) Restoration and Management of Lakes and Reservoirs. 2nd Edition, Lewis Publishers, Boca Raton.

42. Pulatsü, S., Akçora (Topçu), A. and Köksal, F. (2003) Sediment and Water Phosphorus Characteristics in a Pond of Spring Origin, Sakaryabaşı Springs Basin, Turkey. Wetlands, 23, 200-204. http:// dx.doi.org/10.1672/0277-5212(2003)023[0200:SAWPCI]2.0. CO;2

43. Pulatsü, S. and Topçu, A. (2006) Sakaryabaşı Batı Göleti'nde (Türkiye) Sedimentten Fosfor Salınımının Tahmini. Ege Üniversitesi Su Ürünleri Fakültesi Dergisi, 23, 119-121.

44. Topçu, A. and Pulatsü, S. (2014) Phosphorus Fractions and Cycling in the Sediment of a Shallow Eutrophic Pond. Journal of Agricultural Sciences, 20, 63-70.

45. Topçu, A. and Pulatsü, S. (2011) Sakaryabaşı (Çifteler-Eskişehir) Balık Üretim ve Araştırma İstasyonu'nun Su Kaynağı Batı Göleti: Sediment Kaynaklı İnorganik Azot Salınımının Araştırılması. Ekoloji, 20, 26-33. http://dx.doi.org/10.5053/ekoloji.2011.785

46. Topçu, A. and Pulatsu, S. (2008) Phosphorus Fractions in Sediment Profiles of the Eutrophic Lake Mogan, Turkey. Fresenius Environmental Bulletin, 17, 164-172.

47. Pulatsü, S., Topçu, A., Kırkağaç, M. and Köksal, G. (2008) Sediment Phosphorus Characteristics in the Clearwater State

of Lake Mogan, Turkey. Lakes & Reservoirs: Research and Management, 13, 197-205. http://dx.doi.org/10.1111/j.1440-1770.2008.00369.x

48. Pulatsu, S. and Topcu, A. (2009) Seasonal and Vertical Distributions of Porewater Phosphorus and Iron Concentrations in a Macrophyte-Dominated Eutrophic Lake. Journal of Environmental Biology, 30, 801-806.

49. Gürevin, C., Ertürk, A. and Albay, M. (2013) Comparative Study of Nutrient Release from Sediment of the Küçük- çekmece Lagoon under Field and Laboratory Conditions. Proceedings of the 8th Symposium for European Freshwater Sciences, Münster, 1-5 July 2013.

Chapter 5

Applying of Factor Analyses for Determination of Trace Elements Distribution in Water from River Vardar and its Tributaries, Macedonia/Greece

Stanko Ilić Popov[1], Trajče Stafilov[2], Robert Šajn[3],
Claudiu Tănăselia[4], and Katerina Bačeva[1]

[1]RŽ Tehnička Kontrola, 16ta Makedonska Brigada 18, 1000 Skopje, Macedonia

[2]Institute of Chemistry, Faculty of Science, Saints Cyril and Methodius University, 1000 Skopje, Macedonia

[3]Geological Survey of Slovenia, Dimčeva Ulica 14, 1000 Ljubljana, Slovenia

[4]INCDO-INOE 2000, Research Institute for Analytical Instrumentation (ICIA), 67 Donath, 400293 Cluj-Napoca, Romania

ABSTRACT

A systematic study was carried out to investigate the distribution of fifty-six elements in the water samples from river Vardar (Republic of Macedonia and Greece) and its major tributaries. The samples were collected from 27 sampling sites. Analyses were performed by mass spectrometry with inductively coupled plasma (ICP-MS) and atomic emission spectrometry with inductively coupled plasma (ICP-AES). Cluster and R mode factor analysis (FA) was used to identify and characterise element associations and four associations of elements were determined by the method of multivariate statistics. Three factors represent the associations of elements that occur in the river water naturally while Factor 3 represents an anthropogenic association of the elements (Cd, Ga, In, Pb, Re, Tl, Cu, and Zn) introduced in the river waters from the waste waters from the mining and metallurgical activities in the country.

INTRODUCTION

Water is the most essential media for the living world because it supports life processes and without water it would not have been possible to sustain life on Earth. Rivers and streams can be defined as dynamic systems that constantly adjust to natural- and human-caused changes [1]. Generally water resources have a direct influence on the quality of life of the people, their health, and overall productivity. Thus, water is essential, not only to human life but also for animals, agriculture, transport, hydropower generation, industrial development, poverty eradication, and socioeconomic development. Human impacts on the integrity of water resources by altering one or more of five principal factors—physical habitat, seasonal flow of water, the food base of the system, interactions within the biota, and chemical quality of the water [2].

Trace elements, especially anthropogenic elements which mainly consist of heavy metals, have become of particular interest in recent years within the framework of chemical environmental investigations and research. Heavy metals are among the most common environmental pollutants and their occurrence in water and biota indicates presence of natural or anthropogenic sources. The main natural sources of metals in waters are chemical weathering of minerals or soil leaching. A general conclusion and main starting point is that the anthropogenic sources are associated mainly with industrial and domestic effluents, urban storm, water runoff, mining of coal and ore, atmospheric sources, and inputs from the rural areas [3]. It is known that human activities can modify the geochemical cycle of anthropogenic elements resulting in an environmental contamination. Anthropogenic element presence in river water presents a serious threat to aquatic organisms and human life. The determination of trace elements in natural waters is motivated by a number of issues but most importantly because trace elements can play a major role in changing the hydro systems [4].

The water quality and quantity of water resources worldwide is a subject of ongoing concern. During the last few decades, a gradual accumulation of reliable long term water quality data has been monitored for many rivers in the world [5]. Research concerning anthropogenic presence in river water is conducted worldwide. The biggest threat for water systems is discharge of industrial waters, heavy industries, application of fertilizers and pesticides, waste disposals, and so forth [6]. Anthropogenic activity may add considerable amounts of pollution compounds, which will influence the existing aquatic system and change the ecosystem influencing the quality of the aquatic system and treating the aqua life existing in the system [1].

Republic of Macedonia has similar environmental pollution problem with aquatic ecosystems: the developing industry, the agriculture activities, the creation of illegal landfills, and the uncontrolled discharge of faecal waters into rivers contributed to creating a contaminated water ecosystem river named Vardar. As a central water ecosystem river Vardar's basin represents the most important and humanly influenced water resource in the Republic of Macedonia [7–10].

This study deals with determination and interpretation of the presence of trace elements in water samples from various sampling sites of river Vardar and its main tributaries. The goal is to determine the concentration of natural elements that occur in water as well as anthropogenic introduced elements in river Vardar by itself and to determine the concentration of natural and anthropogenic elements that are contributed to Vardar river by its tributaries. By obtaining these results, with interpretation and correlation, a clearer image of anthropogenic presence of various elements in river Vardar and its tributaries will be presented.

MATERIALS AND METHODS

Study Area

In the Republic of Macedonia river Vardar basin (Figure 1) starts form the border with Republic of Kosovo in the north, from the mountain Šar Planina, then from the basins of rivers Lepenec and Južna Morava, to the state border between Republic of Macedonia and Republic of Serbia in the part of the rivers Južna Morava, Pčinja, and Karamanička river. In the east river Vardar basin stretches near the Macedonian-Bulgarian border, with basins of rivers Dvoriska and Strumica, until mountain Belasica and Dojran lake. In the south river Vardar basin stretches near the Macedonian-Greek border in Gevgelija field, with the mountains Kožuf and Nidže, through Pelagonija valley till Baba mountain [11].

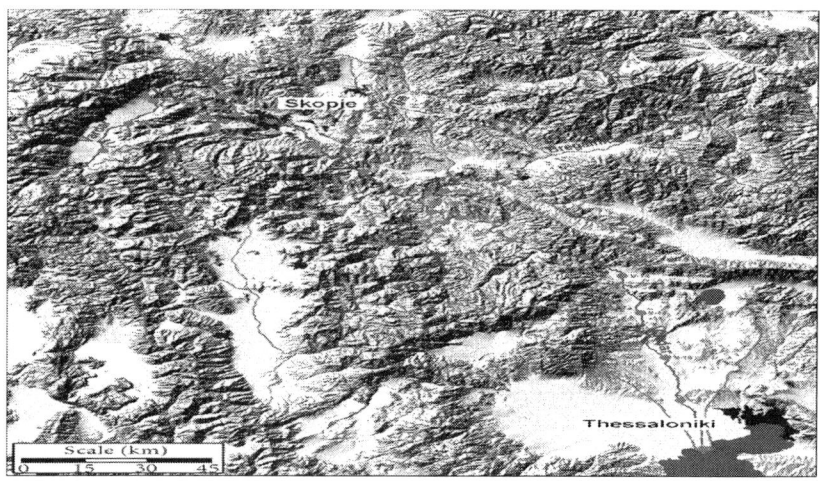

Figure 1: Study area.

River Vardar is the biggest river in the Republic of Macedonia with many tributaries large and small. Main contributors of river Vardar are Treska, Lepenec, Pčinja, Bregalnica, and Crna Reka [12, 13]. The river Vardar catchments area is 24437 km² with 2993 km² (12%) within the Greek territory and 20183 km² in Republic of Macedonia [14, 15].

The climate varies between continental to northern part of the catchments and Mediterranean towards the costal zone. The flow regime is characterized by average flows during the 70s, a wet period during 1980 to 1985, and then dry period from 1986 to 1994. There is a constant decreasing trend in flow between 1980 and 1994 attributed to increased needs for irrigation and drinking water [16]. The watershed topography is characterized by mountainous and semi-mountainous relief, with difference in altitude more than 2800 m which results in high variability in air temperature [13].

The study area (Figure 1) is located from the spring of river Vardar till its discharging into the Aegean Sea. This study includes all the major tributaries of river Vardar (Treska, Lepenec, Pčinja, Bregalnica and Crna River). The study area is located cross the flow of river Vardar and its tributaries. River Vardar occupies 5 valleys and 4 canyons in Republic of Macedonia (Polog valley, canyon Vardarski Derven, Skopje valley, canyon Taor, Veles valley, canyon Veles, Tikves valley, canyon Demir Kapija, and valley Valandovo-Gevgelija) and one canyon in Greece. The length of river Vardar in Republic of Macedonia is 301 km and in Greece 87 km [12]. Treska's river length is 138 km, Lepenec 75 km, Pčinja 135 km, Bregalnica 225 km, and Crna 207 km. River Vardar has average flow of 174 m³/s. The contributes Treska, Lepenec, Pčinja, Bregalnica, and Crna Reka have 30 m³/s, 10 m³/s, 14 m³/s, 28 m³/s, and 37 m³/s average flow consequentially [12].

It is known that the river Vardar and its tributaries are polluted with wastewaters from the municipalities that the river passes through. The presence of illegal landfills with technical and sanitary waste near the river contributes presence of anthropogenic elements. Illegal landfills are present near tributaries as well. Small factories and processing facilities represent potential sources for increasing the anthropogenic elements in the river water. Frequent agricultural activities contribute to polluting the river. Bregalnica, Vardar, and Crna river are polluted from coal mines, Pb-Zn, and Cu mines as well as metallurgical activities [9, 10].

Sampling and Sample Preparation

The water samples were collected in the period June–September 2011 at 28 sites, 18 from the river Vardar starting from the spring finishing at the discharging of river Vardar into the Aegean Sea (Figure 1). Two samples were taken from all tributaries, one sample was taken approximately from 10 to 40 km before discharging into river Vardar and the second sample was taken few kilometers before discharging into river Vardar. Sampling sites from river Vardar were named from V-1 to V-18 (Figure 1). Sampling sites from tributaries were named with the abbreviation of the tributary name, Treska (VTR-1, VTR-2), Lepenec (VLE-1, VLE-2), Pčinja (VPC-1, VPC-2), Bregalnica (VBR-1, VBR-2) and Crna river (VCR-1 and VCR-2). Samples were taken into clean and sterilised plastic bottles of 1 L. Filtration through blue filter paper was made to remove all the organic material from the sample. After filtration 1 mL nitric acid was added to the sample and the sample was preserved.

Instrumentation

The investigated elements Ca, Fe, K, Mg, Na, S, and Si were analyzed by application of inductively coupled plasma atomic emission spectrometry (ICP-AES). The instrument parameters are given in the work of Balabanova et al. [17].

The investigated elements (Al, As, Au, B, Ba, Be, Br, Cd, Ce, Co, Cr, Cs, Cu, Dy, Er, Eu, Ga, Gd, Ho, I, In, La, Lu, Mn, Mo, Nb, Nd, Ni, P, Pb, Pd, Pr, Rb, Re, Rh, Sb, Sc, Sm, Sn, Sr, Tb, Tl, Tm, V, W, Y, Yb, Zn, and Zr) were analyzed by application of mass spectrometry with inductively coupled plasma (ICP-MS). The instrument parameters are given in Table 1.

Table 1: Spectrometer's running parameters

Parameter	Value
Plasma	
Power/W	1350
Plasma gas flow/l min−1	12.00
Auxiliary gas flow/l min−1	1.20
Nebuliser gas flow/l min−1	1.05
Sample/skimmer cone	Platinum
Quadrupole	

Quadruple rod offset (QRO)/V	0.00
Cell rod offset (CRO)/V	−8.00
Cell path voltage (CPV)/V	−20.00
Measurement mode	Peak hopping
Dwell time/ms	Varying
Integration time/ms	Varying
Reading per point	300
Reading per replicate	1
Replicate measurements	4
DRC	
Reaction gas	None
Lens voltage/V	11.00

For all measurements, a SCIEX Perkin Elmer Elan DRC II (Canada) inductively coupled plasma mass spectrometer (with quadrupole and single detector) was used. The operating parameters are listed in Table1. The instrument's running parameters were checked and adjusted every day of measurements, using a solution with 1 ppb In, 1 ppb Ce, 10 ppb Ba, and 1 ppb Th and Mg. Oxides levels and double ionized levels were kept under 3%, background for both low and high mass was under 1 cps, and all the other parameters were chosen considering the best signal/noise ratio. The dynamic reaction chamber (DRC) was used in RF-only mode (no gas) and its parameters optimization have been given earlier [18]. For sample introduction system, a classic setup was used, consisting in a peristaltic pump, a Meinhard nebulizer, and a cyclonic spray chamber, where the fine aerosols are formed that goes directly into the plasma. 18 MΩ cm^{-1} DI water was prepared in the laboratory using a Millpore-Milli-Q ultra pure water purification system. All measurements were done using the semi quantitative method (TotalQuant) supplied by Elan 3.4 software that uses a response factor calibration curve which was obtained by a calibration in multiple points, low, medium, and high concentration, for optimum setup, using a multielement Merck VI standard solution diluted to mimic real sample consumption.

RESULTS AND DISCUSSION

The descriptive statistics presented in Table 2 shows the results of total 69 elements in all 28 water samples of river Vardar and its tributaries. Table 2 shows the results and the parameters for the analysed elements, number of samples, distribution, unit in which concentration of elements is expressed, arithmetic mean, geometric mean, median, minimum, maximum, 25 percentiles, and 75 percentiles.

Table 2: Descriptive statistics of chemical analysis of river water from Vardar and its tributaries

Element	N	Dis.	Unit	X	Xg	Md	Min	Max	P25	P75
Al	27	Log	μg/L	0.20	0.12	0.12	0.02	1.3	0.05	0.25
As	27	Log	μg/L	2.4	1.8	1.9	0.24	6.9	1.1	3.3
Au	27	Log	ng/L	16	2.2	0.50	0.50	126	0.50	11
B	27	Log	μg/L	5.8	4.4	4.2	0.44	17	3.4	7.5
Ba	27	Log	μg/L	28	24	22	9.6	97	18	29
Be	27	Log	ng/L	34	17	12	5.0	283	5.0	31
Br	27	Log	μg/L	29	20	21	3.0	117	12	36
Ca	27	Log	mg/L	53	53	52	40	70	46	60
Cd	27	Log	ng/L	83	28	22	5.0	510	5.0	65
Ce	27	Log	μg/L	0.99	0.37	0.38	0.05	10	0.11	0.74
Co	27	Log	μg/L	0.72	0.55	0.57	0.18	4.0	0.34	0.76
Cr	27	Log	μg/L	1.11	0.06	0.01	0.01	7.3	0.01	1.3
Cs	27	Log	ng/L	26	9.7	15	0.50	214	3.0	31
Cu	27	Log	μg/L	4.8	3.4	3.0	1.3	17	2.1	6.4
Dy	27	Log	ng/L	86	24	25	1.0	986	6.0	80
Er	27	Log	ng/L	38	11	8.0	1.0	423	4.0	37
Eu	27	Log	ng/L	29	15	14	3.0	235	7.0	26
Fe	27	Log	mg/L	0.31	0.15	0.14	0.03	2.3	0.06	0.30
Ga	27	Log	ng/L	154	33	41	5.0	1133	5.0	105
Gd	27	Log	ng/L	120	35	41	5.0	1315	5.0	95
Ho	27	Log	ng/L	16	5.4	4.0	0.50	183	2.0	15
I	27	Log	μg/L	0.28	0.13	0.12	0.01	1.0	0.05	0.35
In	27	Log	ng/L	106	2.3	0.50	0.50	1040	0.50	3.0
K	27	Log	mg/L	4.0	3.4	3.4	0.76	10	2.1	4.6
La	27	Log	μg/L	0.43	0.17	0.18	0.03	4.2	0.05	0.36
Lu	27	Log	ng/L	4.8	1.9	1.0	0.50	51	1.0	5.0

Element	N	Dis.	Unit	X	X_g	Md	Min	Max	P25	P75
Mg	27	N	mg/L	11	10	10	4.5	16	9.4	12
Mn	27	Log	µg/L	103	28	43	1.3	1204	5.3	73
Mo	27	Log	µg/L	0.31	0.23	0.25	0.03	0.81	0.12	0.45
Na	27	Log	mg/L	17	13	15	2.6	54	8.1	17
Nb	27	Log	ng/L	8.8	7.2	5.0	5.0	29	5.0	11
Nd	27	Log	µg/L	0.45	0.15	0.13	0.01	4.7	0.05	0.37
Ni	27	Log	µg/L	2.5	1.7	1.6	0.43	14	1.1	2.0
P	27	Log	µg/L	146	115	122	10	524	97	188
Pb	27	Log	µg/L	6.4	1.4	1.5	0.05	73	0.61	4.2
Pd	27	Log	ng/L	1.5	1.0	0.50	0.50	6.0	0.50	3.0
Pr	27	Log	ng/L	109	37	34	5.0	1122	11	91
Rb	27	Log	µg/L	1.4	1.2	1.1	0.27	3.8	0.80	1.8
Re	27	Log	ng/L	3.2	1.8	1.0	0.50	14	1.0	4.0
Rh	27	Log	ng/L	16	12	10	4.0	59	8.0	19
S	27	Log	mg/L	7.4	6.4	6.4	1.7	19	4.2	9.4
Sb	27	Log	µg/L	1.2	0.68	0.47	0.30	6.1	0.37	0.70
Sc	27	Log	µg/L	2.2	2.0	1.9	1.1	4.1	1.6	2.6
Si	27	Log	mg/L	5.8	5.6	5.4	3.3	11	4.3	7.0
Sm	27	Log	ng/L	976	33	28	5.0	969	13	104
Sn	27	Log	ng/L	15	12	13	5.0	41	5.0	16
Sr	27	Log	µg/L	136	122	116	65	347	88	157
Tb	27	Log	ng/L	16	5.2	5.0	0.50	178	1.0	17
Tl	27	Log	ng/L	46	3.4	0.50	0.50	404	0.50	28
Tm	27	Log	ng/L	5.7	2.1	2.0	0.50	59	1.0	4.0
V	27	Log	µg/L	1.3	0.08	0.01	0.01	6.5	0.01	2.2
W	27	Log	ng/L	6.6	3.3	4.0	0.50	28	1.0	9.0
Y	27	Log	µg/L	0.42	0.15	0.15	0.03	4.5	0.04	0.36
Yb	27	Log	ng/L	32	12	5.0	5.0	356	5.0	28
Zn	27	Log	µg/L	13	8.2	6.7	1.3	95	5.6	13
Zr	27	Log	ng/L	50	41	42	12	114	24	750

Element: analysed element; N: number of samples analysed; Dis.: distribution; Unit: unit in which concentration of elements is expressed; X: arithmetic mean; X_g: geometric mean; Md: median; Min: minimum; Max: maximum; P25: 25 percentiles; P75: 75 percentiles.

There is a lot of industrial activity by river Vardar and its basin: chemical industry, metallurgy, cements industry, smelter factories, fertilizer factories, factories for food and drink production, and so forth. All these industrial activities have negative influence on river Vardar and its tributaries. The cities which river Vardar passes through do not have plants for wastewater treatments so the wastewater flows directly into river Vardar without proper treatment. Anthropogenic influence is determined from agricultural areas because of usage of phosphate fertilizers which contain Cd. Wastewaters from factories flow into river Vardar untreated as well. Heavy metals occur in river Vardar because of Pb-Zn smelter plant in city of Veles, steel production including galvanized steel from iron-steel factory in Skopje, coals and heavy oil used for energy needs for Toplifikacija and Skopski Leguri in Skopje, chemical industry Ohis in Skopje, and fertilizer plant near Veles [9, 10].

Table 2 shows the descriptive statistics of chemical analysis of river water from Vardar and its tributaries. Twenty-seven samples were analyzed and results for 56 element concentrations were obtained. As presented the distribution of all elements in the table is logarithmic with the exception of Mg where the distribution is normal.

Seven analyzed elements have concentrations in range of mg/L (Ca, Fe, K, Mg, Na, S, and Si). These elements occur in river waters naturally and are considered a natural part of the river water. Depending on the concentration of anthropogenic elements present in the river, elements that occur naturally in water like Si or S can have highly variable concentrations [3].

Twenty-three analyzed elements (Al, As, B, Ba, Br, Ce, Co, Cr, Cu, Ga, I, La, Mn, Mo, Nd, Ni, P, Pb, Rb, Sb, Sc, Sr, V, Y, and Zn) have concentrations in range of µg/L. Knowing that some anthropogenic elements (Al, As, Co, Cr, Cu, Mn, Ni, and Pb) are found in wastewaters coming from some industrial facilities [9, 19] into river Vardar it is not clear that these elements are present in these results as well. It is found that the range of the concentration of these elements is between 3 and 280 µg/L [20] just in Skopje area. Higher concentration of Pb in some of the water samples is as a result of the pollution from the activities of the smelter plant for lead and zinc in Veles [19].

As expected some of the examined elements have very low concentrations. Total sum of 25 elements have concentrations in the range of ng/L. The presence of twenty-four out of twenty-five elements in these concentrations is considered normal for river waters because these elements do not occur in river water or in pollutants. Cd is an anthropogenic element that is present in the examined samples in concentrations of ng/L but its presence in river water is due to the pollution from wastewaters from the Pb-Zn smelter plant in Veles [19].

The highest mean, geometric mean, median, minimum, maximum, 25 percentiles, and 75 percentiles values were obtained from Ca with 53 mg/L, 53 mg/L, 52 mg/L, 40 mg/L, 70 mg/L, 46 mg/L, and 60 mg/L consequentially. The lowest average value was obtained from Rb, 1.4 ng/L. Lowest geometric mean value was obtained from V (0.08 ng/L). Lowest median was obtained from Au, In, Pd, and Tl showing the concentration of 0.5 ng/L. Au, Cs, Ho, In, Lu, Pd, Tb, Tl, Tm, and W showed lowest minimum value of 0.5 ng/L. Lowest maximum value showed was from Pd with 6 ng/L. Lowest P_{25} value showed was from Au, In, Pd, and Tl with 0.5 ng/L. Lowest P_{75} value showed was from In and Pd with 3 ng/L. In general, the presence of anthropogenic elements is due to the presence of different types of industries near the rivers. Main contributors of anthropogenic elements in river waters are the ferro-alloys plant "Silmak" near Jagunovce (former ferro-chromium smelter plant) for Cr; the factories in the main city Skopje for Al, As, Co, Cr, Cu, Mn, Ni, and Pb; next, the smelter plant in Veles for Pb, Zn, and Cd; Mines Sasa and Zletovo in the water catchment of river Bregalnica for contribution of Pb, Zn, and Cd in the river waters; and Ferronickel ore processing and smelter plant near city Kavadarci for Fe and Ni contribution into the river system of Crna river [13, 21, 22].

Factor analysis is the basic of statistical techniques that are used to analyze relations of numerous variables. The goal is to process numerous information gained from original variables and turn it into smaller one (factors) with minimal loss of information from the original variables. Factor analysis has been made using Statistica 6.1 program. Multivariate factor analysis with R-method is applied displaying the association of the chemical elements. For orthogonal projection varimax method is used with signification 0.5 and four factors have been obtained (F1, F2, F3, and F4), Table 3.

Table 3: Factor analysis

Element	Fac	F1	F2	F3	F4
As	1	0.92	−0.01	0.08	0.17
B	1	0.92	−0.06	−0.01	0.20
Br	1	0.95	−0.09	−0.03	−0.18
K	1	0.98	0.07	−0.02	−0.11
Mg	1	0.93	0.03	0.18	−0.04

Mo	1	0.74	0.03	0.47	0.02
Na	1	0.98	0.01	0.03	0.01
Rb	1	0.97	0.17	0.01	−0.07
Rh	1	0.89	0.22	−0.05	0.12
S	1	0.95	0.05	0.16	−0.14
Sr	1	0.92	0.22	0.06	−0.16
Al	2	−0.15	0.93	0.03	0.09
Ba	2	0.36	0.80	0.15	0.32
Be	2	0.05	0.76	−0.08	0.35
Co	2	0.49	0.66	0.37	−0.07
Cs	2	0.43	0.52	−0.49	0.06
Cu	2	0.15	0.71	0.18	−0.32
Fe	2	−0.06	0.96	−0.14	0.06
Mn	2	−0.01	0.94	−0.21	0.00
Ni	2	0.52	0.75	−0.01	0.27
P	2	0.22	0.69	0.22	0.15
Sn	2	0.23	0.67	0.07	−0.12
Y	2	−0.05	0.95	−0.15	0.12
Zn	2	0.01	0.87	0.24	−0.08
Zr	2	−0.23	0.71	0.18	−0.01
La_Lu	2	0.17	0.77	0.11	0.26
Cd	3	0.18	0.30	0.83	−0.12
Ga	3	0.08	0.30	0.87	0.04
In	3	0.05	−0.11	0.96	0.05
Pb	3	−0.20	0.60	0.71	0.03
Re	3	0.27	0.12	0.88	0.10
Sb	3	−0.12	−0.33	0.83	0.32
Tl	3	0.13	−0.01	0.85	0.24
Ca	4	0.58	0.27	−0.03	−0.68
Sc	4	0.02	0.41	0.21	0.77
Si	4	−0.13	0.34	0.16	0.84
W	4	0.51	0.02	0.33	0.59
Prp. Totl		30.41	28.37	16.82	8.14
Expl. Var		11.25	10.50	6.22	3.01
Eigen V.		13.57	8.86	5.96	2.60

Factor 1 has the highest variance factor and represents a mixed group of elements. Factor 1 is the strongest factor and represents 30.41% from the total variability. Factor one represents a group of elements naturally found in river water: As, B, Br, K, Mg, Mo, Na, Rb, Rh, S, Sr, and Ni. Elements Ca and W will be explained in Factor 4 because of stronger correlation. Presence of these elements is detected in every sample taken from river Vardar and all tributaries with exception of spring sample. The delta has the highest values because these elements are more present in sea water then river water. All tributaries contribute approximately with same concentrations of the elements from this factor to river Vardar. K, Mg, Na, and Ca are elements that also occur naturally in river water. Concentration of B in river Vardar and its tributaries is low. Concentrations of Mg, Mo, and Nb in river Vardar and its tributaries is equally without any drastic changes. Conclusion can be made that either there is constant contamination with these elements or that they occur naturally in the water.

The results for the factor scores are presented in the histograms (Figure 2) with two parts: Vardar-dist (left histogram) and main tributaries (right histogram). Vardar-dist section is divided in 7 parts along the river: spring (V-1), average value for the samples from 20th to 60th km from the spring with the samples V-2, V-3, and V-4 (from city Gostivar to village Jagunovce), from 80 to 130 km (V-5, V-6, and V-7) from village Jagunovce to exit of city Skopje, from 150 to 180 km (V-8, V-9, V-10) from exit of city Skopje to exit of city Veles, from 190 to 260 km (V-11, V-12, V-13, and V-14) from village Dubrovo to Macedonian-Greek border, from 280 to 360 km (V-15, V-16, and V-17) from Macedonian-Greek border to Delta, and the water sample from the delta (V-18).

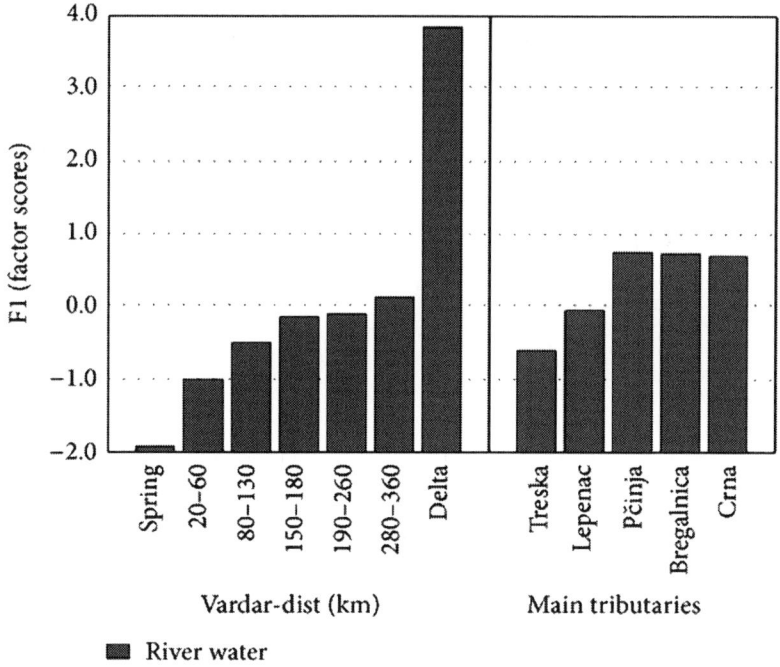

Figure 2: Factor 1 (factor scores) of river Vardar and main tributaries.

Factor 1 represents 30.41% from the total variance including As, B, Br, Ca, K, Mg, Mo, Na, Ni, Rb, Rh, S, Sr, and W. Figure 2 shows that the biggest factor scores for Factor 1 are observed from the sample from the delta. This is normal because the sample taken from the delta is mostly salty water. Salty water has bigger concentration of the examined elements gained in Factor 1. Results from the delta aside, higher factor scores are gained in the tributaries of river Vardar-Pčinja, Bregalnica, and Crna Reka.

According to the Decree for categorization of rivers, lakes, accumulations, and groundwaters of the Republic of Macedonia [23], most of river Vardar and tributaries is considered second and third class water. According to the Decree for water classification of the Republic of Macedonia [24] there are no limitations for presence of the elements in river water that are gained in Factor 1, except for As, Ni, and Mo (Decree for water classification, 1999) [24]. Concentrations found in samples taken from river Vardar and tributaries of As, Ni, and Mo have maximum values of 6.9 µg/L, 14 µg/L, and 0.81 µg/L which are concentrations far below the permissible limits (30 µg/L and 500 µg/L, resp.). Nevertheless, As is an anthropogenic element if found in bigger concentrations. As shown in Figure 3 the concentration of As is rising consequentially from the spring to the delta. Tributaries Pčinja and Crna Reka show higher concentrations of As in comparison to other tributaries which is mostly as a consequence of geological origin [25].

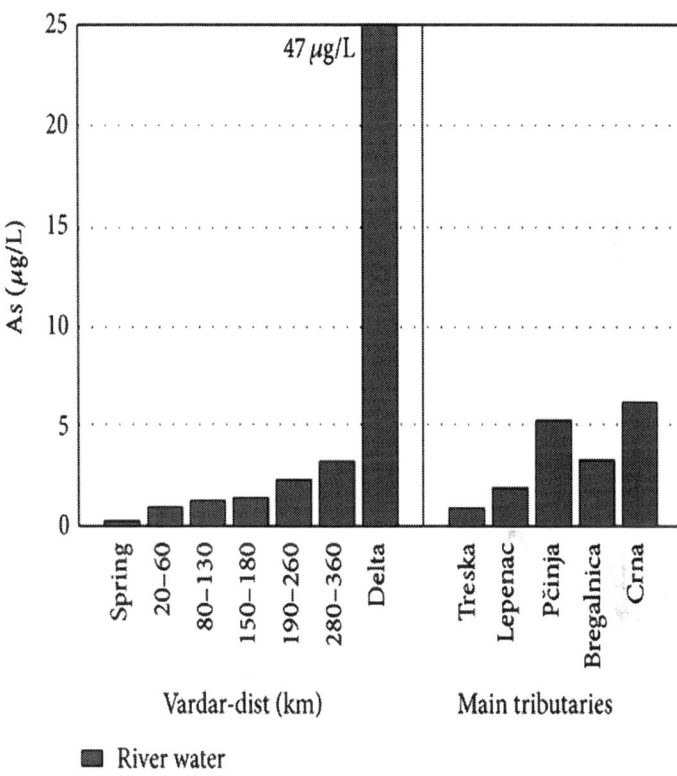

Figure 3: Concentration of As found in river Vardar and main tributaries.

Factor 2 represents 28.37% from the total variance. Factor 2 is represented by Al, Ba, Be, Co, Cs, Cu, Fe, Mn, Ni, P, Sn, Y, Zn, Zr, La, and Lu. These elements are usually found naturally in river water in low concentrations and most of them present essential elements for life forms but in higher concentrations represent threat to human life, biota, and the environment.

The factor scores of Factor 2 are presented in Figure 4 for separate lengths of river Vardar and the tributaries. The spring of river Vardar shows lower factor scores in comparison to factor scores from 20 to 60 km and 80 to 130 km. A small decrease in the factor score is present in the region 150–180 km and then a decreasing trend is present till the delta. Highest factor scores are presented in tributary Bregalnica and then tributaries Lepenec and Pčinja. The reason for the increased factor scores for this part of river Vardar and river Bregalnica is due to the anthropogenic influence by Cu, Zn, and P due to Pb-Zn and Cu mining activities in this region as well as due to the agricultural activities and use of phosphate fertilizers. In Figures 5–7 the concentrations of Cu, Zn, and P found in river Vardar and main tributaries are presented.

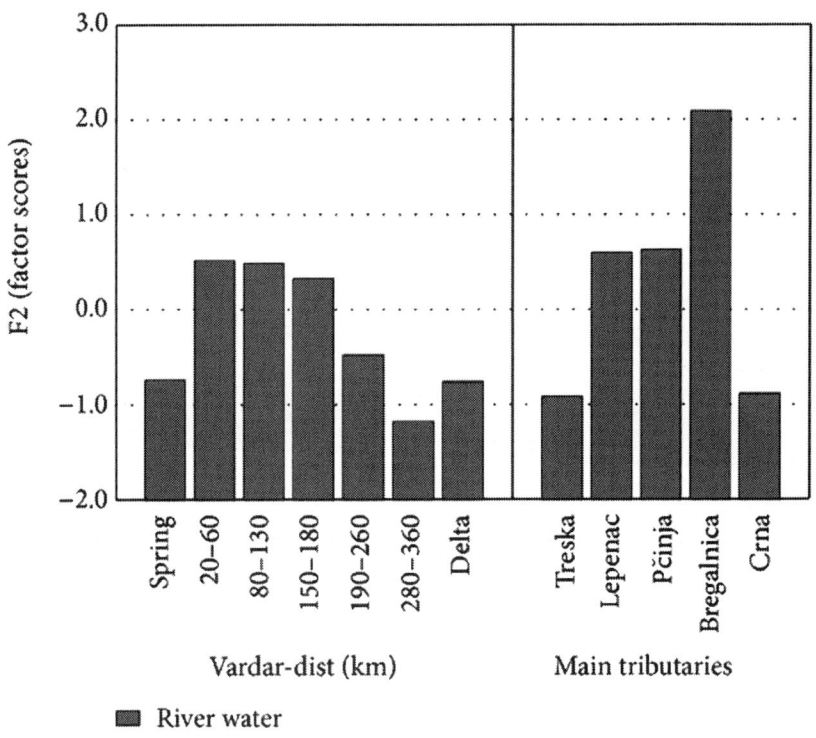

Figure 4: Factor 2 (factor scores) of river Vardar and main tributaries.

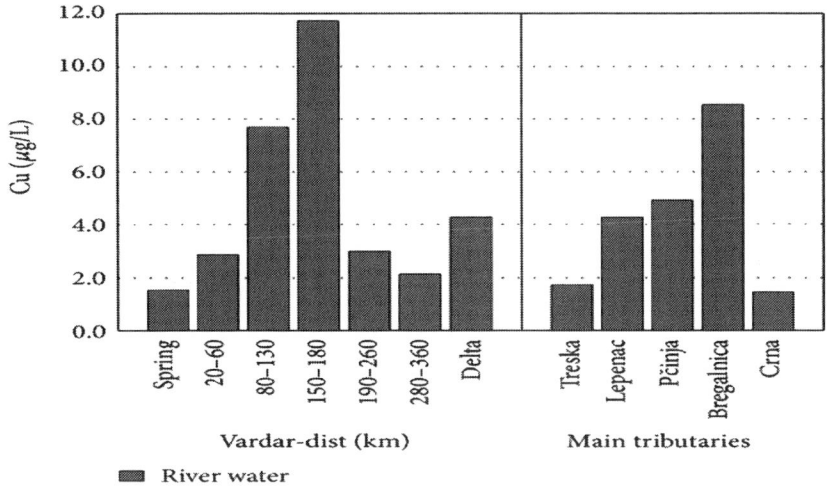

Figure 5: Concentration of Cu found in river Vardar and main tributaries.

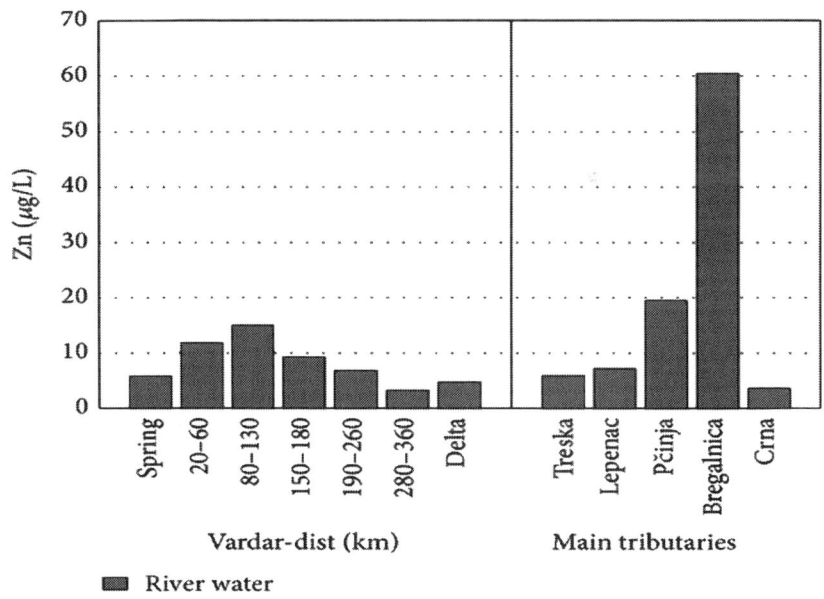

Figure 6: Concentration of Zn found in river Vardar and main tributaries.

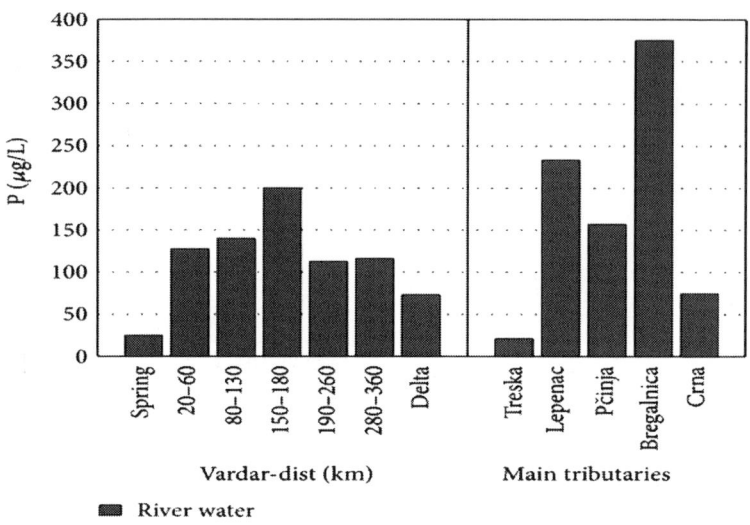

Figure 7: Concentration of P found in river Vardar and main tributaries.

Higher concentrations of Cu close to the MAC (10 µg/L) are determined in parts of river Vardar (in samples from 80–130 and 150–180 km) and Bregalnica tributary (8.5 µg/L, 11.5 µg/L, and 8.4 µg/L, resp.). In 1989 it was found that concentrations of Cu in river Vardar was 16.4 ppb [26]. It was found that wastewaters that flow into the Vardar river in the area of Skopje have concentration of Cu between 70 and 91 ppm [20]. The trend of disposing wastewaters into river Vardar showed higher concentrations in water samples taken after the city of Skopje. Tributary to higher concentrations of Cu in river Vardar in samples taken after city Skopje are domestic wastewaters as well. Higher concentrations of Cu are determined in water samples taken after city Veles. The Pb-Zn smelter in city of Veles influences the water quality of river Vardar. Namely, according to Stafilov et al. (2008, 2010) [19, 27] the average amount of Cu in topsoil in city of Veles ranges between 11 and 1700 mg/kg.

Figure 6 represents a typical example of environmental pollution of river waters with Zn. As shown, in comparison to all other sampling sites river Bregalnica has the highest values for the concentration of Zn of 60 µg/L. As mentioned, the presence of three mines near river Bregalnica contributes to higher concentrations of Zn in the river water [9, 10].

The waters of tributary Bregalnica have higher concentrations of Cu and Zn. Many wastewaters are discharged into river Bregalnica from flotation processes in Bučim mine and Sasa and Zletovo mines [10, 17,28–30]. Wastewater from flotation processes spread up to Vardar river and Aegean Sea. The wastewaters from overburden leaching are extremely concentrated and contain up to 840 mg/L Cu and 360 mg/L Mn, in an average flow of 2 L/s [31].

Similar to river Vardar, river Sitnica in Kosovo is a recipient of wastewaters coming from industrial plants and agriculture, without any previous treatment or purification process. River Sitnica, similar river Vardar, has several tributaries that receive wastewaters from plants and agriculture and contribute anthropogenic elements to river Sitnica when they discharge. Cu and Zn are determined into samples in both research. River Vardar showed maximum presence of copper of 95 μg/L in comparison to river Sitnica with 2.5 mg/L. River Vardar showed maximum presence of 17 μg/L, and river Sitnica showed 60 mg/L [32]. It can be concluded that both rivers are contaminated with Cu and Zn, but river Vardar is less contaminated in comparison to river Sitnica. The average flow of river Vardar is 174 m³/s and river Sitnica has an average flow of 0.8 m³/s. Heavy metals have the tendency to accumulate in river beds and river sediments. Because of the greater flow of river Vardar it is possible for the heavy metals to be more accumulated in the sediments in comparison to the water itself.

Higher concentrations of P are determined in samples taken from tributaries Lepenec and Bregalnica because of developed agriculture and fertilizing in areas where Lepenec and Bregalnica flow. Relatively high concentrations of total phosphorus are observed in the Vardar river indicating pollution mainly from phosphorus detergents and fertilizers [13].

Factor 3 represents 16.82% from the total variance. Elements that are part of this factor are with anthropogenic character (Figure 8). The anthropogenic group consists of Cd, Ga, In, Pb, Re, Tl, Cu, and Zn. The median value for Cd is 22 ng/L and the maximum 510 ng/L. The median value for Ga is 41 ng/L and the maximum 1133 ng/L. The median value for In is 0.5 ng/L and the maximum 1040 ng/L. The median values for Pb and Sb are 1.5 μg/L and 0.47 μg/L successive and maximum values are 73 μg/L and 6.1 μg/L. The median values for Cu and Zn are 3.0 μg/L and 42 μg/L successive and maximum values are 17 μg/L and 114 μg/L.

The median values for Re and Tl are 1 ng/L and 0.5 ng/L successive and maximum values are 14 ng/L and 404 ng/L.

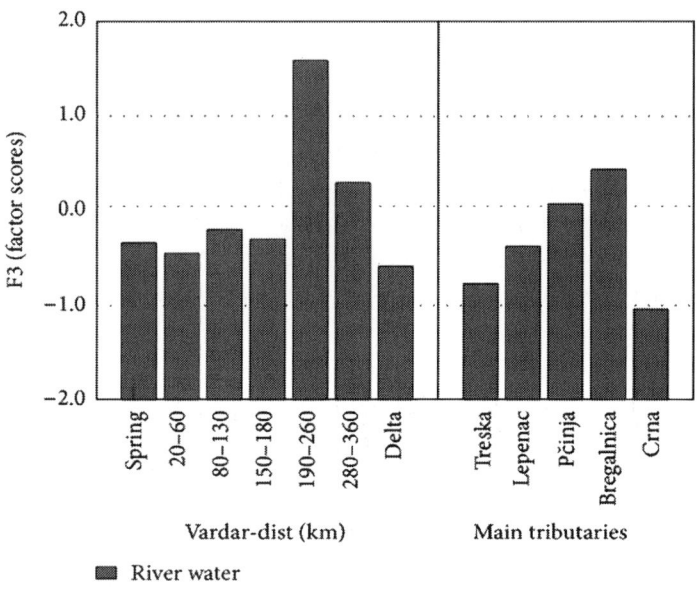

Figure 8: Factor 3 (factor scores) of river Vardar and main tributaries.

As presented in Figure 8, the factor scores for the anthropogenic elements are constant in river Vardar with exception of two points 190–260 km and 280–360 km. This is normal because samples from the point 190–260 km are collected after all the tributaries are into river Vardar. Normally the factor scores will be higher in the samples taken from 280 to 360 km because the river Vardar still carries the anthropogenic elements with its flow. All tributaries show higher factor scores for anthropogenic elements with the exception of Crna river where the factor scores are very low.

Higher concentrations of Cd were determined in samples taken after city Veles and tributary Bregalnica (Figure 9). This is expected; because of presence of Pb-Zn smelter factory in Veles, higher concentrations occur in river Vardar. The Pb-Zn mines Sasa and Zletovo contribute anthropogenic elements to tributary Bregalnica [10, 19]. It is known that soil in the city of Veles and environs has content of Cd between 0.3 and 600 mg/kg, content of Pb 13 to 1500 mg/kg, and content of Sb 0.016

to 105 mg/kg [19, 27]. Knowing this fact it is expected to determine higher concentrations of these elements in water samples taken from river Vardar after the city of Veles. Furthermore, concentrations of 1.12 mg/L for Pb and 0.03 mg/L were determined in Bregalnica [21]. Identical to river Vardar, river Kamchia in Bulgaria is polluted with anthropogenic elements from smelter plants, industrial activities, and untreated wastewaters. Unlike river Vardar with contribution of 73 µg/L of Pb to the Aegean Sea, river Kamchia has 115 t/year contribution of Pb into the Black Sea. The difference occurs because of greater mining and industrial activities near river Kamchia. The contribution of Cd into the Aegean and Black seas is present from both river flows. River Vardar contributes 0.51 ng/L to the Aegean Sea and river Kamchia contributes approximately 10 t/year into the Black Sea [33].

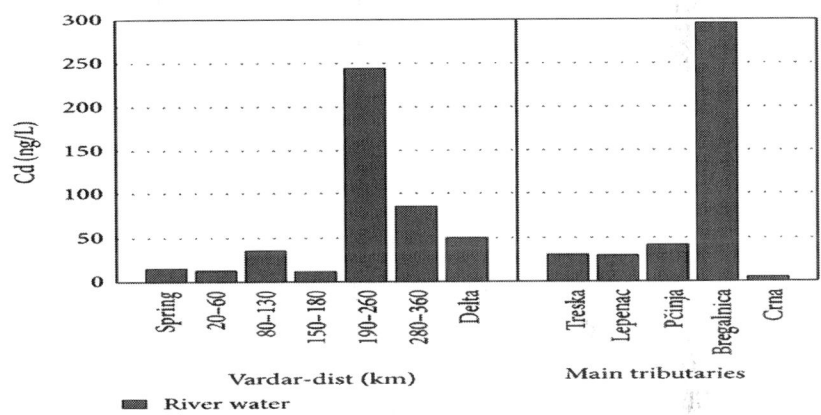

Figure 9: Concentration of Cd found in river Vardar and main tributaries.

Figure 10 shows the concentration of Pb in river Vardar and contributes. Again, high concentration of Pb is present after 150 km because of Pb/Zn smelter factory in Veles. Samples taken from Bregalnica tributary like mentioned showed higher concentrations of Pb because Pb-Zn mines Sasa and Zletovo [10]. Similar to river Vardar, river Sava in the Republic of Serbia showed presence of Cd and Pb in the water flow. In comparison to the results gained from water samples from river Vardar, river Sava showed greater concentrations for Pb and smaller concentrations for Cd (73 µg/L in comparison to 7.2 µg/L for Pb, 510 ng/L in comparison to 4.1 µg/L for Cd) [34].

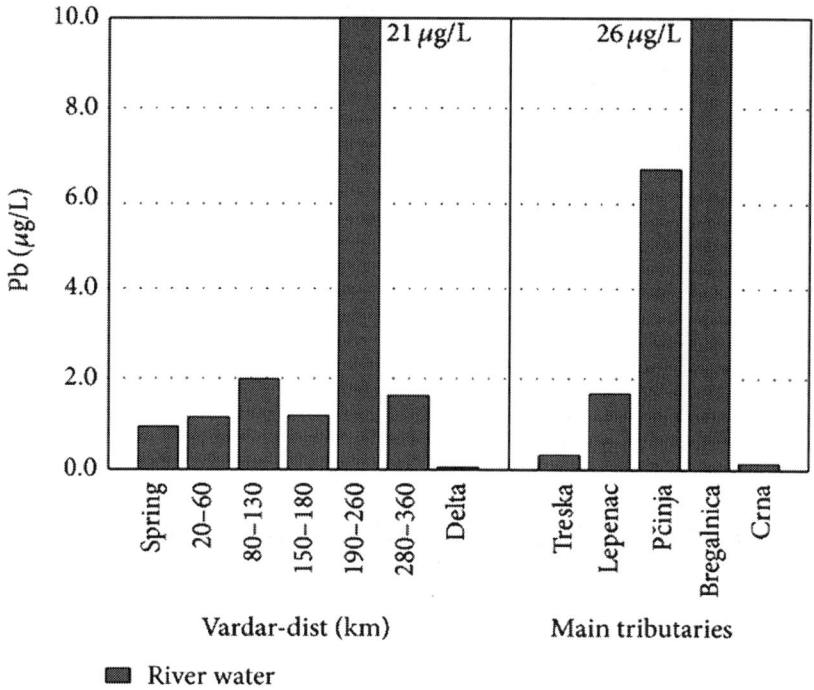

Figure 10: Concentration of Pb found in river Vardar and main tributaries.

To summarize and compare all results present in Factor 1, 2, and 3 (whiteout P) an observation can be made between results gained from water samples from river Vardar and its tributaries and results from mine waters in north-western Bulgaria [35]. Normally, higher concentrations can be found into mine waters in comparison to river waters.

Factor 4 (Ca, Si, Sc, and W) represents 8.14% from the total variance. Elements that are part of this factor occur naturally in the river water. Figure 11 presents Factor 4 (factor scores) of river Vardar and main tributaries. Factor four is represented by Ca, Sc, Si, and W. These are elements that occur naturally in river water. Concentrations of Ca are almost the same in all the samples collected from river Vardar and tributaries except in delta where concentration of Ca is much higher in comparison with all the other samples. This is understandable because in the delta, river Vardar mixes with salt water from the Aegean Sea. Determined concentrations of Si have the similar story as concentrations of Ca in the samples. Concentrations of Si are almost the same in all the samples collected from river Vardar and tributaries except in delta where concentration of Si is much lower. Determined concentrations of Sc and W in river samples are very low. It is known that concentrations of Sb and Sn in river Vardar are far below the permissible levels [36].

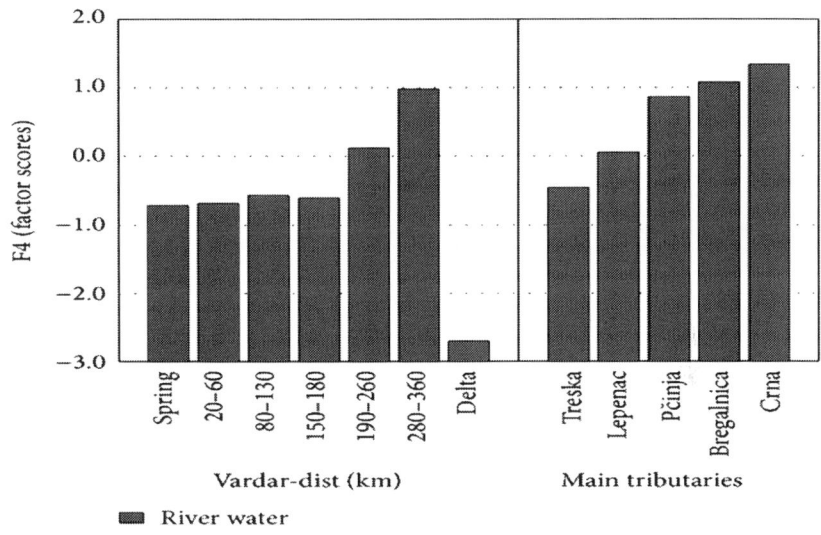

Figure 11: Factor 4 (factor scores) of river Vardar and main tributaries.

It is interesting to mention the distribution of Cr in the river water. Although not a part of any factor presented in this research the distribution of Cr gives an interesting behavior. Figure 12 represents the distribution of chromium in river Vardar and main tributaries. Concentrations of chromium are very low until 80 km from the spring. After 80 km concentrations of Cr are increasing rapidly. This is because of pollution from the slag waste deposit from the former ferrochromium smelter plant, from the mining activities of former chromium mine in Radusa region [10], and from the water of river Pčinja where the concentrations of Cr is much higher that in the waters from the other tributaries. Also 80 km from the spring starts the main city Skopje and afterwards the city of Veles, both known by their industrial activities.

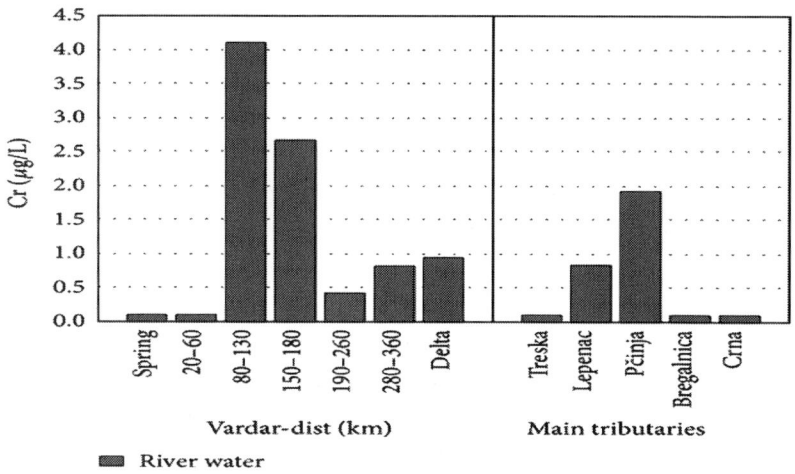

Figure 12: Concentration of Cr found in river Vardar and main tributaries.

CONCLUSIONS

Fifty-six elements in river water samples of river Vardar and its tributaries were analyzed. A sum of 28 sampling sites in Republic of Macedonia and Greece was established. Descriptive statistics was made showing greater concentrations of anthropogenic elements in river Vardar and tributaries. Factor analysis was made and 4 factors gained. Three factors were consisted of elements that occur naturally in the river water and one factor consisted of anthropogenic elements. Although Cu and Zn are elements that are part of Factor 2 higher concentrations of these are present in the water which is toxic to the environment and dangerous to human health. Three factors represent the associations of elements that occur in the river water naturally while Factor 3 represents an anthropogenic association of the elements. The anthropogenic factor, Factor 3, showed correlation between Cd, Ga, In, Pb, Re, Sb, and Tl. All these elements present in river Vardar and its tributaries represent a group of potentially threatening elements for human health and the environment. All these elements are found in river waters because of industrial activities.

REFERENCES

1. R. J. Naiman and R. E. Bilby, River Ecology and Management, Wayerhaevser Company, Tacoma, Wash, USA, 2008.

2. R. M. Bhardwaj, J. Chilton, J. van der Gun et al., Water Qquality for Ecosystem and Human Helath, GEMS Water Programme, Ontario, Canada, 2nd edition, 2008.

3. A. Kabata-Pendias and A. B. Mukherjee, Trace Elements form Soil to Human, Springer, Berlin, Germany, 2007.

4. J. I. Drever, Surface and Ground Water, Weathering and Soils, Elsevier, Amsterdam, The Netherlands, 1st edition, 2005.

5. H. El-Shaarawi and R. E. Kwiatkowski, Statistical Aspects of Water Quality Monitoring, Elsevier, Amsterdam, The Netherlands, 1986.

6. S. Darby and D. Sear, River Restoration—Managing the Uncertainty in Restoring Physical Habitat, John Wiley & Sons, Chichester, UK, 2004.

7. S. Krstić, L. Melovski, Z. Levkov, and P. Stojanovski, "Complex investigations on the river the river Vardar. II: the most polluted sites in the first three months," Ecololy and Protection of the Environment, vol. 2, no. 2, pp. 13–29, 1994.

8. Z. Levkov and S. Krstic, "Use of algae for monitoring of heavy metals in the River Vardar, Macedonia," Mediterranean Marine Science, vol. 3, no. 1, pp. 99–112, 2002.

9. T. Stafilov and Z. Levkov, Summary of Vardar River Basin Field Survey, European Agency for Reconstruction and Ministry of Environment & Physical Planning of the Republic of Macedonia, Skopje, Macedonia, 2007.

10. T. Stafilov, L. Peeva, B. Nikov, and A. de Koning, "Industrial hazardous waste in the Republic of Macedonia," in Applied Environmental Geochemistry—Anthropogenic Impact on Human Environment in the SE Europe, R. Šajn, G. Žilbert, and J. Alijagić, Eds., pp. 108–112, Geological Survey of Slovenia, Ljubljana, Slovenia, 2009.

11. Water Economy Master Plan of R. Macedonia, Ministry of Water Economy, Forestry and Agriculture, Skopje, Macedonia, 1974.

12. M. B. Gasevski, Waters of Macedonia, Skopje, Macedonia, 1972.

13. M. Milovanovic, "Water quality assessment and determination of pollution sources along the Axios/Vardar River, Southeastern Europe," Desalination, vol. 213, no. 1–3, pp. 159–173, 2007.

14. A. P. Karageorgis, N. P. Nikolaidis, H. Karamanos, and N. Skoulikidis, "Water and sediment quality assessment of the Axios River and its coastal environment," Continental Shelf Research, vol. 23, no. 17–19, pp. 1929–1944, 2003.

15. M. J. Diamantopoulou, V. Z. Antouopoulos, and D. M. Papamichail, "The use of a neural network technique for the prediction of water quality parameters of Axios river in Northern Greece,"European Water, vol. 11, no. 12, pp. 55–62, 2005.

16. P. Karageorgis, Ed., General's Description of the Axious River Catchments and the Gulf of Thermaikos, Catchment Changes and Their Impact on the Coast, European Catchments, Athens, Greece, 2001,http://www.iia.cnr.it/big_file/EUROCAT/publications/EUROCAT%20WD07.pdf.

17. B. Balabanova, T. Stafilov, R. Šajn, and K. Bačeva, "Characterisation of heavy metals in lichen species Hypogymnia Physodes and Evernia Prunastri due to biomonitoring of air pollution in the vicinity of copper mine," International Journal of Environmental Research, vol. 6, no. 3, pp. 779–792, 2012.

18. C. Tănăselia, M. Micelan, C. Roman, E. Cordoş, and L. David, "Determination of lead isotopic ratio in organic and soil materials using a Quadrupole mass spectrometry method with fast inductively coupled plasma," Optoelectronics and Advanced Materials, vol. 2, pp. 229–302, 2008.

19. T. Stafilov, R. Šajn, Z. Pančevski, B. Boev, M. V. Frontasyeva, and L. P. Strelkova, "Heavy metal contamination of topsoils around a lead and zinc smelter in the Republic of Macedonia," Journal of Hazardous Materials, vol. 175, no. 1–3, pp. 896–914, 2010.

20. S. Lepitkova and B. Boev, "Heavy and toxic metals in the waste waters from some industrial facilities in the town of Skopje, Republic of Macedonia," Geologica Macedonica, vol. 13, pp. 97–104, 1999.

21. Zendelska, M. Golomoeva, B. Krstev, B. Golomeov, A. Krstev, and Z. Panov, "The impact of tailing dam of Sasa minion the

quality of surrounding waters," in Proceedings of the 14th Balkan Mineral Processing Congress, vol. 2, pp. 677–680, Tuzla, Bosnia and Herzegovina, June 2011.

22.   K. Bačeva, T. Stafilov, and R. Šajn, "Biomoniotring of nickel air pollution near the city of Kavadarci, Republic of Macedonia," Ecology and Environmental Protection, vol. 12, no. 1-2, pp. 57–69, 2009.

23.   "Decree for categorization of rivers, lakes, accumulations and groundwaters," Official Journal of Republic of Macedonia, no. 18, pp. 1173–1179, 1999.

24.   "Decree for water classification," Official Journal of Republic of Macedonia, no. 18, pp. 1165–1173, 1999.

25.   T. Stafilov, R. Šajn, and J. Alijagić, "Distribution of arsenic, antimony and thallium in soil in Kavadarci and its environs, Republic of Macedonia," Soil and Sediment Contamination, vol. 22, no. 1, pp. 105–118, 2013.

26.   N. T. Skoulikidis, I. Bertahas, and T. Koussouris, "The environmental state of freshwater resources in Greece (rivers and lakes)," Environmental Geology, vol. 36, no. 1-2, pp. 1–17, 1998.

27.   T. Stafilov, R. Sajn, Z. Pancevski, B. Boev, M. V. Frontasyeva, and L. P. Strelkova, Geochemical Atlas of Veles and the Environs, Faculty of natural Sciences and Mathematics, Skopje, Macedonia, 2008.

28.   B. Balabanova, T. Stafilov, K. Bačeva, and R. Šajn, "Biomonitoring of atmospheric pollution with heavy metals in the copper mine vicinity located near Radovis, Republic of Macedonia," Journal of Environmental Science and Health A, vol. 45, pp. 1504–1518, 2010.

29.   B. Balabanova, T. Stafilov, R. Šajn, and K. Bačeva, "Distribution of chemical elements in attic dust as reflection of their geogenic and anthropogenic sources in the vicinity of the copper mine and flotation plant," Archives of Environmental Contamination and Toxicology, vol. 61, no. 2, pp. 173–184, 2011.

30.   B. Balabanova, T. Stafilov, R. Šajn, and K. Bačeva, "Comparison of response of moss, lichens and attic dust to geology and atmospheric pollution from copper mine," International Journal of Environmental Science and Technology, 2013.

31. S. Hadži Jordanov, M. Maletić, A. Dimitrov, D. Slavkov, and P. Paunović, "Waste waters from copper ores mining/flotation in "Bučbim" mine: characterization and remediation," Desaniation, vol. 213, no. 1–3, pp. 65–71, 2007.

32. T. Arbneshi, M. Rugova, and L. Berisha, "The level concentration of lead, cadmium, copper, zinc and phenols in the river Sitnica," Journal of International Environmental Application & Science, vol. 3, no. 2, pp. 66–72, 2008.

33. S. I. Dineva, "Water discharges into the Bulgarian Black Sea," in Proceedings of the International Symposium on Outfall Systems, Mar del Plata, Argentina, May 2011,http://www.osmgp.gov.ar/symposium2011/Papers/78_Dineva.pdf.

34. Ž. Vuković, M. Radenković, S. J. Stanković, and D. Vuković, "Distribution and accumulation of heavy metals in the water and sediments of the River Sava," Journal of the Serbian Chemical Society, vol. 76, no. 5, pp. 795–803, 2011.

35. D. Dimitrova, Z. Cholakova, N. Velitchkova et al., "Heavy meral and metalloid concentration dynamics in mine and surface waters in the vicinity of the Chiprovtsi and Martinovo mines, Northwestern Bulgaria," Bulletin of Geological Society of Greece, vol. 40, no. 3, pp. 1397–1408, 2007.

36. J. M. Serafimovska, S. Arpadjan, T. Stafilov, and S. I. Popov, "Dissolved inorganic antimony, selenium and tin species in water samples from various sampling sites of river Vardar in Macedonia and Greece," Macedonian Journal of Chemistry and Chemical Engineering, vol. 30, no. 2, pp. 181–188, 2011.

Trace Metal Concentration in Two Matrices in an Urban Subtropical River

Nyasha Mabika[1], Trust Masiya[2], Beaven Utete[3], Maxwell Barson[4], and Joshua Tsamba[4]

[1]Department of Anatomy, University of Zimbabwe, Harare, Zimbabwe
[2]Institute of Mining Research, University of Zimbabwe, Harare, Zimbabwe
[3]Department of Wildlife Ecology and Conservation, Chinhoyi University of Technology, Chinhoyi, Zimbabwe

ABSTRACT

This study investigates the concentration of metals namely aluminium, manganese and cobalt in two matrices: sediment and fish organs (whole muscle stomach tissue, gills, liver and kidney) in an urban river, Mukuvisi River, Zimbabwe. River bed sediments and fish samples were collected simultaneously at five sites over seven months (September

2008-April 2009). Concentrations of aluminium, manganese and cobalt in the selected fish organs and sediment were estimated using the Flame Atomic Absorption Spectrometry (FAAS). Water limnochemical aspects, dissolved oxygen, pH, temperature and conductivity were measured concomitantly at each site. Aluminium had significantly higher mean concentrations and bioconcentration factors in both sediments and fish tissues relative to cobalt and manganese. Cobalt and aluminium were detected in all fish tissues, whilst manganese was not detected in muscle and liver. Significant differences in bioconcentration factors for the metals in organs of the same fish species analysed in this study show differences in metal assimilation. Metal specific river rehabilitation methods need to be applied for the future restoration of the ecological integrity of Mukuvisi River.

INTRODUCTION

The chemistry of metals in sediments and water in rivers with respect to bio-availability to aquatic organisms and chemical reactivity is regulated by pH, texture, and organic matter contents, specific binding form and coupled reactivity of the metals [1]. Metals accumulate through the aquatic chain and their bioavailability is an indirect measure of potential toxicity [2]. Faced with rapid industrialization and an accelerating demotechnic growth, urban authorities deliberately discharge sewage and industrial effluent into urban streams [3]. This effluent contains metals which have potential deleterious impacts on the aquatic fauna [4]. In natural freshwater ecosystems most metals are typically present in low concentrations such that any increase in the levels of bio- available metals in the sediment and water phases may lead to an increase in the bioaccumulation of the metals in tissues of aquatic organisms [5] [6] .

Among the aquatic fauna, fish are most susceptible to heavy metal toxicants and are thus more vulnerable to heavy metal contamination than any other aquatic fauna [3] . When fish are exposed to elevated concentrations of metals in a polluted aquatic ecosystem, they tend to absorb these metals directly from their environments [7] [8]. After uptake, the metals are transported by the blood, where they are usually bound to proteins and brought into contact with the tissues and consequently accumulate in different tissues [9] [10]. The accumulated heavy metals may ultimately reach concentrations hundreds or thousands of times above those measured in the water, sediment and food [11].

The Mukuvisi River, drain the greater part of the city of Harare, the most industrialized and urbanized city in Zimbabwe [12]. As a result, the river displays a gradient in pollution as most of the urban, agricultural and industrial run off uploads into it before it uploads into Lake Chivero Because of the threat of water pollution, several studies to monitor the water quality of Mukuvisi River have been done. However, most of these studies have focused on monitoring of the river integrity using resident aquatic organisms particularly macroinvertebrates [13] [14] , selected water quality parameters [15] - [18] , metazoan fish parasites [19] and selected heavy metals [20] , largely neglecting the essential and trace metal elements like manganese, aluminium and cobalt analysed in this study. The main objectives of this study were to investigate the concentration of aluminium, manganese and cobalt in the sediment phase and assess their potential bioavailability to the African catfish (Clarias gariepinus, Burchell 1822) in an urban river, Mukuvisi River, Zimbabwe.

MATERIAL AND METHODS

Sampling Sites

Five sampling sites were selected along the Mukuvisi River. The Seke Road Bridge was the first site. At this site, industrial effluent (from Graniteside and City Centre) and leachate from landfills are presumed to be discharged into the river. Site 2 was the Magaba bend industry. Effluent is expected to be discharged from the informal industry in the Magaba area into the river. Site 3 was Glen Norah where most of the effluent from the Southert on industry is discharged into the river. Site 4 was the Amalinda farm. Effluent from urban cultivation such as fertilizers and pesticides are leached into the river. Site 5 was pension farm where effluent from the Firle Sewage Treatment Plant is presumed to be discharged into the river. The sampled sites in the Mukuvisi River are shown in Figure 1.

Sample Collection

Sediment and fish were sampled at five selected sites in the Mukuvisi River once after every two months (from September 2008 to April 2009). Sampling occurred between 1000 and 1400 hours on all occasions to minimize variation due to climate factors. Sediment samples were collected using an acid-washed polyethene corer to a depth of 3 cm. At each sampling site three replicate samples were collected at an interval distance of 15 m. The sediments were stored in sterilized polyethene bags and returned to the laboratory in sealed containers containing ice packs for metal analysis. Live fish samples were collected at each site using 89 mm meshfyke nets and a DC electro fisher (Smith-Root Type IV-A). Electro fishing was done for 10 minutes at each sampling station.

Water Quality

Temperature, pH, conductivity and dissolved oxygen (DO) were measured in situ. Water temperature and DO were measured using a HACH oxygen 330i meter; pH and conductivity were measured using a HACH pH meter and WTW 330i conductivity meter, respectively. Water samples were analysed in the laboratory for total and reactive phosphorus (orthophosphate), total nitrogen, nitrate and ammonia. The nutrients were measured with a HACH water analysis kit (HACH DR/2010 portable data logging spectrophotometer) using filtered water samples filtered through 0.45 μm Whatman GF 47 mm filters, except for total nitrogen and phosphorus analysis. The methods are described in detail in [21].

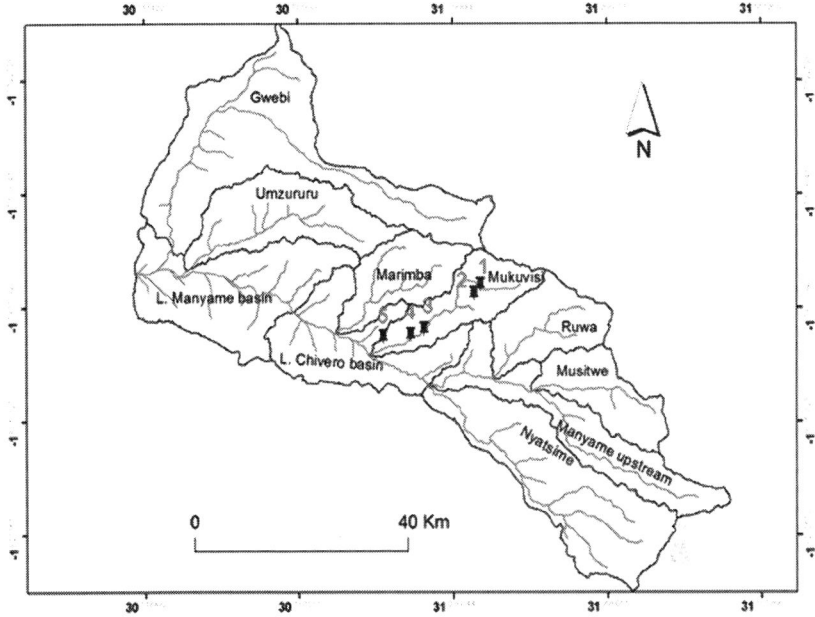

Figure 1: Location of sampling sites in the Mukuvisi River.

Heavy Metals in Sediments

The two step acid digestion process used for the sediments is a strong digestion method and was chosen so that the total metal content could be estimated. Metals in the sediments were analysed using flame atomic absorption spectrophotometry (FAAS) [22] after acid digestion to extract the metals from the sediments. This procedure consisted of two parts:

- Digestion: The sediments were oven-dried at 180°C in a muffle furnace and large aggregates were broken up and then 20 ml nitric acid and 5 ml perchloric acid were added to 5 g of the oven-dried sample. The mixture was heated on a hot plate until fumes were produced and allowed to cool to room temperature.

- Acidification: 20 ml of 50% hydrochloric acid was added to the mixture from the first digestion. The acidified mixture was heated until boiling and then was cooled to room temperature. The acidified mixture was filtered and distilled water was added to the filtrate in a volumetric flask up to the 100 ml mark. The digested sediments were then analysed for metal amounts in an atomic absorption spectrometer [22].

Heavy Metals in Fish Tissues

In the laboratory, fish were dissected with clean autoclaved stain less teel instruments. One gram (wet weight) of gill, liver, kidney and muscle tissue of C. gariepinus was measured using an electronic balance. The tissues were placed into Petri dishes to dry at 120°C until they reached a constant weight. Drygill, liver, kidney and muscle tissue were placed in digestion flasks to which 5 ml perchloric acid and 10 ml nitric acid were added. All acids used were of analytical grade quality. The digestion flasks were placed in an oven at 130°C until all materials dissolved. The prepared tissue samples were analysed using flame atomic absorption spectrophotometry (FAAS).

Data Analysis

Spatial differences in limnochemical aspects of the water among sites sampled in the Mukuvisi River were investigated using Kruskall Wallis ANOVA at 5% level of significance respectively using Paleontological Statistics (PAST) softwareVersion 190 [23]. The bioconcentration factor (BCF) was calculated after metal concentrations were determined for sediment and fish tissues. The BCF is defined in this study as the concentration of a metal in a fish tissue in relation to the concentration of that metal in the sediment surrounding that tissue, and was calculated using the formula [24]:

$$BCF = \frac{\text{Concentration of metal in fish organ (ppm)}}{\text{Concentration of metal in river sediments (ppm)}}$$

BCF values greater than 1000 are considered high and those under 250 low. Those between these extremes are classified as moderate. Differences in bioconcentration factors were assessed using the Kruskall Anova at 5% significance level.

Canonical correspondence of the metal concentration in the sediment and fish tissue to water limnochemistry was analysed using CANOCCO 5 [25] .

RESULTS

Limnochemistry of the Mukuvisi River

The pH of the water at all sites sampled in the Mukuvisi River was slightly acidic to alkaline (6.1 - 7.3) during the period of study, only site 1 had a pH below the lower limit accepted for effluent and river water (Table 1). The mean temperature varied with the time of sampling ranging from 20.8°C to 22.1°C. Conductivity readings were above 540 $\mu S \cdot cm^{-1}$ at all sites. Dissolved oxygen was highest at site 2 (5.0 mg·l^{-1}) and lowest at site 5(1.1 mg·l^{-1}). Phosphate levels exceeded the WHO limits (0.5 mg·l^{-1}), whilst nitrate levels remained within the acceptable limits of 10.0 mg·l^{-1}. The highest levels of ammonia were recorded at site 3 (0.18 mg·l^{-1}) whilst the lowest levels were recorded at site 1 (0.03 mg·l^{-1}). There were significant differences (Kruskall ANOVA; $p < 0.05$) in the concentration of D.O, RP and TP among sites sampled in the Mukuvisi River.

Metal Concentration in Mukuvisi River Bed Sediment and Catfish Tissues

All the three metals (Al, Mn and Co) were recorded in sediments at all sites (Table 2). Aluminium had the highest concentrations in the sediments followed by manganese. The levels of aluminium ranged from 17.20 - 215.00 mg·kg^{-1} in sediments. Manganese levels ranged from1.44 - 72.40 mg·kg^{-1}, whilst the levels of cobalt ranged from 0.20 - 2.85 mg·kg^{-1}. No fish were collected from sites 1 and 5 (Table 2). Aluminium was not detected in gills and kidney on site 3, whilst manganese was not detected in fish muscle and liver. Manganese was not detected in the kidney on sites 2 and 3. Cobalt was not detected in the liver on sites 2 and 3. Aluminium ranges in the fish tissues were; muscle (0.27 - 1.18 µg·g^{-1}), gill (ND-1.35 µg·g^{-1}), kidney (ND-2.7

$\mu g \cdot g^{-1}$), liver (0.06 - 1.13 $\mu g \cdot g^{-1}$). Manganese ranges in the fish tissues were; gill (0.15 - 0.85 $\mu g \cdot g^{-1}$), kidney (0.10 - 0.30 $\mu g \cdot g^{-1}$), whilst cobalt ranges were; muscle (0.01 - 0.1 $\mu g \cdot g^{-1}$), gill (0.02 - 0.13 $\mu g \cdot g^{-1}$), kidney (0.01 - 0.08 $\mu g \cdot g^{-1}$) and liver (0.03 - 0.05 $\mu g \cdot g^{-1}$). The prominence (hierarchical ranking) order of mean concentration of metals in the sediments was; Al > Mn > Co. Prominence order of metals in the muscles and liver of the catfish was; Al > Co >Mn, whilst in the gills and kidneys it was; Al >Mn> Co (Table 3).

Metal accumulation pattern (bioconcentration factors) differed for each organ analysed and comprised of the following pattern: Aluminium: Liver > Gill > Muscle > Kidney, Manganese: Kidney > Gill > Liver > Muscle and for Cobalt: Liver > Muscle = Kidney > Gill (Table 4). Aluminium had the highest bioconcentration factors in all organs of the catfish analysed except in the kidney where Mn had a significantly higher bioconcentration (Table 4). There was a significant difference (ANOVA, $p < 0.05$) in the bioconcentration the three metals from the sediments into the catfish stomach muscles and liver.

Table 1: Water quality parameters in sites sampled in the Mukuvisi River. DO = Dissolved oxygen, RP = Reactive phosphorous, TP = Total phosphorous, TN = Total nitrogen. EMA and WHO [35] threshold levels are highlighted

Site	pH	Temp °C	Conductivity μS·cm−1	DO mg·l−1	RP mg·l−1	TP	TN mg·l−1	Nitrate mg·l−1	Ammonia mg·l−1
1	6.1	20.8	553.3	3.9	2.10	6.36	61.00	0.53	0.03
2	6.9	20.9	578.0	5.0	3.60	2.41	83.66	0.61	0.11
3	7.1.1	22.1	603.0	2.2	1.00	3.20	42.74	0.69	0.18
4	7.3	22.1	544.0	3.7	8.21	5.21	52.06	0.42	0.08
5	7.3	22.1	623.0	1.1	4.62	6.78	45.57	0.37	0.12
WHO	6.59	25.30	1000	5.0	5.0	10.0	5.0	200	5.0
EMA	6.59	25.30	1000	5.0	5.0	10.0	5.0	200	5.0

Table 2: Ranges, mean ± SD of the concentrations of Al, Mn and Co in sediments (mg·kg⁻¹/ppm) and tissues (µg·g⁻¹/ppm) of C. gariepinus

Metal	Site	Sediments	Muscle	Gill	Kidney	Liver
	1	Rs (74.00 - 142.00) M (110.9 ± 27.64)	No fish	No fish	No fish	No fish
	2	Rs(78.00 - 118.70) M (95.97 ± 20.91)	Rs (0.40 - 0.50) M (0.45 ± 0.05)	Rs (0.92 - 1.30) M (1.10 ± 0.19)	Rs (0.80 - 1.51) M (1.20 ± 0.39)	Rs (0.40 - 1.00) M (0.70 ± 0.30)
Al	3	Rs (62.60 - 102.00) M (84.10 ± 16.79	Rs (0.80 - 1.18) M (0.99 ± 0.19)	ND	ND	Rs (0.06 - 0.15) M (0.10 ± 0.04)
	4	Rs (17.20 - 215.00) M(117.9-98.00)	Rs (0.27 - 0.74) M (0.50 ± 0.23)	Rs (0.39 - 1.35)) M (0.83 ± 0.48)	Rs (1.45 - 2.7) M (2.02 ± 0.63)	Rs (0.84 - 1.13) M (1.00 ± 0.15)
	5	Rs (43.50 - 70.00) M (58.7 ± 13.67	No fish	No fish	No fish	No fish
	1	Rs (18.42 - 72.40) M (31.02 ± 28.96)	No fish	No fish	No fish	No fish

	2	Rs (13.69 - 13.90) M (13.80 ± 0.15)	ND	Rs (0.43 - 0.85) M (0.64 ± 0.21)	ND	ND
Mn	3	Rs (2.63 - 11.69) M (7.36 ± 3.74)	ND	Rs (0.25 - 0.41) M (0.33 ± 0.08)	ND	ND
	4	Rs (1.44 - 10.57) M (3.74 ± 4.76)	ND	Rs (0.15 - 0.50) M (0.32 ± 0.17)	Rs (0.10 - 0.30) M (0.14 ± 0.05)	ND
	5	Rs (3.47 - 6.30) M (5.25 ± 1.55)	No fish	No fish	No fish	No fish
	1	Rs (0.82 - 1.78) M (1.16 ± 0.54)	No fish	No fish	No fish	No fish
	2	Rs (0.50 - 2.85) M (1.29 ± 1.10)	Rs (0.01 - 0.1) M (0.05 ± 0.04)	Rs (0.06 - 0.13) M (0.09 ± 0.04)	Rs (0.03 - 0.05) M (0.04 ± 0.01)	ND

Co	3	Rs (0.28 - 0.72) M (0.51 ± 0.18)	Rs (0.02 - 0.03) M (0.03 ± 0.01)	Rs (0.02 - 0.08) M (0.05 ± 0.03)	Rs (0.01 - 0.03) M (0.01 ± 0.02)	ND
	4	Rs (0.21 - 1.28) M (0.66 ± 0.53)	Rs (0.02 - 0.04) M (0.03 ± 0.01)	Rs (0.04 - 0.09) M (0.07 ± 0.03)	Rs (0.06 - 0.08) M (0.07 ± 0.01)	Rs (0.03 - 0.05) M (0.04 ± 0.010
	5	Rs (0.20 - 0.43) M (0.33 ± 0.12)	No fish	No fish	No fish	No fish

Note: ND: not detected, Rs: ranges, M: mean.

Table 3: Summary of mean ± SD concentrations of metals in sediments (ppm or mg·kg⁻¹) and fish tissues (ppm or µg·g⁻¹)

Metal	Sediment	Muscle	Liver	Gill	Kidney
Mn	12.23 ± 11.18	ND	ND	0.43 ± 0.18	0.14 ± 0.00
Al	93.25 ± 23.30	0.65 ± 0.30	0.60 ± 0.46	0.64 ± 0.57	1.27 ± 0.71
Co	0.78 ± 0.42	0.04 ± 0.01	0.01 ± 0.02	0.07 ± 0.02	0.05 ± 0.03

Note: ND: Not detected.

Table 4: Bioconcentration factors of Al, Mn and Co from sediments to selected catfish tissues in Mukuvisi River

Metal	Fish organ			
	Muscle	Liver	Gill	Kidney
Mn	12.23	12.23	28.44	87.36
Al	143.50	155	145.7	73.43
Co	19.5	78	11.14	19.5

Canonical correspondence analyses show that there was a strong association in aluminum concentration in the kidney, liver and gill of the catfish with the levels of DO and TN in water in Mukuvisi River particularly at site 2, whilst there is a strong association of Al concentration in the catfish muscles with nitrate levels in the water at Mukuvisi site 1. The first two axes explain 87.61% of the variation in Al concentration in the catfish tissues sampled (Figure 2). A narrow angle of attachment shows that temperature; conductivity and ammonia values were closely linked in the water phase in Mukuvisi River. The correspondence analysis for manganese was not valid as the number of ND non-detectable values in the catfish tissues which were treated as missing figures or zeros exceeded the possible explanatory variables (water limnochemical parameters). Figure 3 shows that the cobalt concentration in the muscle and gill of the catfish had a strong association with the nitrate, ammonia and DO values in the Mukuvisi River. Cobalt concentration in the kidney and liver of the catfish sampled in the Mukuvisi River had a strong association with the RP values in the water. The first two axes explain only 64.13% of the variation in the concentration of Co concentration in the catfish tissues analysed in this study. A narrow angle of attachment shows that temperature, conductivity and TN values are closely related in the water phase in Mukuvisi River (Figure 3).

DISCUSSION

The objectives of this study were to investigate the concentration of Al, Co and Mn in two matrices: sediments and selected fish tissues of the African catfish Clarias gariepinus, Burchell 1822, in a subtropical urban river, Mukuvisi River, and assess potential bioavailability of metals to the fish. Results show Al, Co and Mn contamination in sediments and fish tissues. This contamination is attributed to localized pollution sources along the Mukuvisi River mainly automobile and detergent manufacturing industries which directly discharge metal laden effluent into the river [13] [20] . Al, Co and Mn concentrations in sediments were significantly higher relative to fish tissues. This is expected as sediments act both as a source and a sink for metals [26] . However, an elevated concentration of metals in sediments translates to increased potential bioavailability to resident aquatic fauna [27] . For instance, aluminium which has highest concentration in sediments among the three metals we analysed has a higher bioconcentration factor in all the fish tissues except in the kidneys where Mn had the highest bioconcentration factors. An increased bioavailability of a metal is not problematic as long as the lethal dose limit is not exceeded [28] .

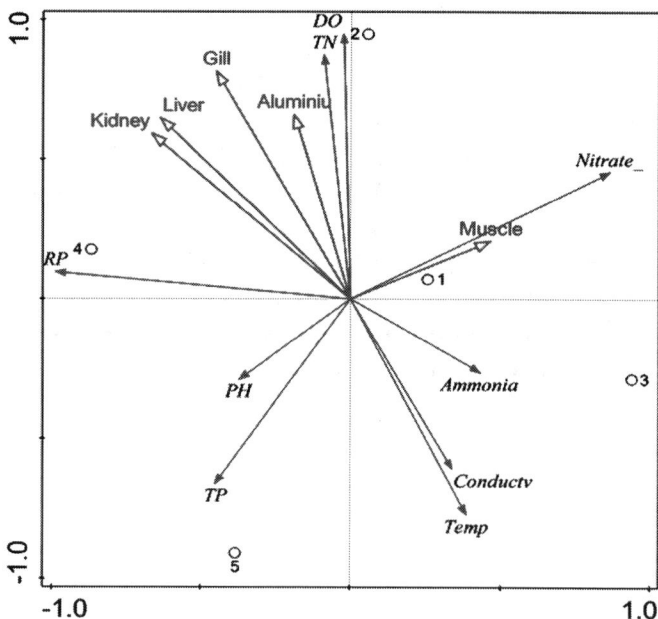

Figure 2: Correspondence analyses of the Al concentration in catfish tissues and selected water limnochemical parameters in sites sampled in the Mukuvisi River.

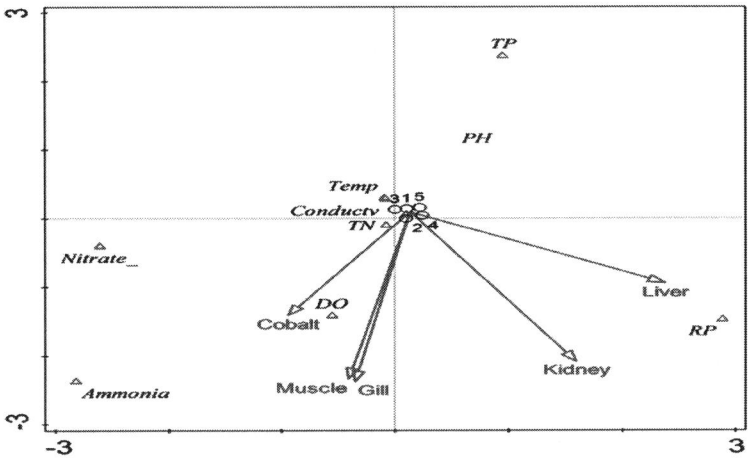

Figure 3: Correspondence analyses of the Co concentration in catfish tissues and selected water limnochemical parameters in sites sampled in the Mukuvisi River.

Significant differences in bioconcentration factors for the metals in organs of the same fish species analysed in this study show differences in metal assimilation. Metal assimilation in an aquatic organism corroborates closely with its specific binding form (species/fraction), chemical reactivity, and depuration [29] . Aluminium has a range of species but the most assimilated fractions by freshwater aquatic organisms are the Al^{2+} and Al^{3+} ions [30] . Speciation modelling of the assimilation and toxicity of Al to freshwater fish has been correlated to the water limnochemistry particularly the concentration of dissolved organic carbon, dissolved oxygen levels and pH [30] [31] . In this study correspondence analysis shows that Al levels in the sediment and fish tissue matrices is strongly associated with the concentrations of dissolved oxygen (DO), total nitrogen (TN) and nitratesin the water phase in the Mukuvisi River.

Manganese is a toxic element frequently overlooked when assessing the toxicity of sewage and industrial effluents, sediments and pore waters [32] [33] . Manganese exists in different fractions but the most assimilated form is the Mn^{2+} ion in freshwater [32] . Several authors have correlated the rapid assimilation and acute toxicity of Mn^{2+} ion in freshwater organisms to stochastic shifts in pH, as well as DO concentration. In well aerated waters with slightly alkaline pH (pH > 7), Mn^{2+} concentration is low in lotic systems due to redox equilibrium reactions that favour its conversion to the less toxic Mn (IV) fraction which is easily sequestered into insoluble Mn oxides [32] [33] . Manganese analyses show non-detectable (ND) incidences at most sites in both the sediment and fish matrices in the Mukuvisi River. This could be due to the rapid conversion of the highly reactive and radical Mn^{2+} ion into the insoluble Mn (IV) fraction which is assimilated by aquatic organisms as it is an essential element. Hence we could not investigate the possible relation between Mn concentration in the sediments and fish tissues to the levels of water limnochemistry parameters.

Elevated concentrations of Co were observed in the sediments mainly at Mukuvisi site 1 and in the liver of the catfish. The elevated concentrations of cobalt at site 1 can be attributed to the industrial effluent discharge from Graniteside Industrial Area and City Centre and leachate from adjacent landfills [20] . Cobalt is an essential metal vital for flora and fauna as it forms the backbone of the cofactors in most vitamins. However, elevated concentrations in aquatic systems can be detrimental to the fauna as elevated cobalt (particularly Co^{2+} ion) toxicity has been demonstrated to lungs, heart and skin of amphibians and fish [34] . Toxicity and assimilation of cobalt have been correlated to water pH and water hardness [34] . For our study the cobalt concentration in the sediment and the fish tissues (kidney and liver) of the catfish sampled in the Mukuvisi River had a strong association with the RP values in the water.

Results from this study demonstrate metal (Al, Co and Mn) contamination in the sediment and fish tissue matrices in the Mukuvisi River. Furthermore there is accumulation/bioconcentration of Al, Co and Mn in catfish tissues thus indicating their potential toxicity. In addition our analyses show that there is possible association and relation of Al and Co concentration in sediments and catfish tissues to the stochastic levels of water limnochemical parameters in the Mukuvisi River. Although the lack of metal fraction analyses and the specific sediment toxicity tests underestimate the risks associated with each metal, use of total components in the analysis of metal contamination in sediment and fish tissues provides a useful river integrity assessment option for management. In future studies there is need for both single and multispecies sediment toxicity tests for any specified metal, and metal specific river rehabilitation methods need to be applied for the rehabilitation and restoration of the Mukuvisi River.

ACKNOWLEDGEMENTS

The laboratory and field assistance of Elizabeth Munyoro is greatly appreciated.

REFERENCES

1. Harichandan, R., Routroy, S., Mohanty, J.K. and Randa, C.R. (2013) An Assessment of Heavy Metal Contamination in Soils of Freshwater Aquifer System and Evaluation of Ecotoxicityby Lithogenic Implications. Environmental Monitoring Assessment, 185, 3503-3516. http://dx.doi.org/10.1007/s10661-012-2806-7

2. Storelli, M.M., Storelli, A., Diddaabbo, R., Morano, C., Bruno, R. and Marcotrigiano, G.O. (2005) Trace Elements in Loggerhead Turtles (Caretta caretta) from the Eastern Mediterranean Sea: Overview and Evaluation. Environmental Pollution, 135, 163-170.http://dx.doi.org/10.1016/j.envpol.2004.09.005

3. Agbozu, I.E., Ekweozor, I.K.E. and Opuene, K. (2000) Survey of Heavy Metals in the Catfish (Synodontisclarias). International Journal of Environmental Science Technology, 4, 93-97.

4. Obire, O., Tamauro, P.C. and Wemedo, S.A. (2003) Physico-Chemical Quality of ElechiCreek in Port Harcourt, Nigeria. Journal of Applied Science and Environmental Management, 7, 43-50.

5. Du Preez, H.H., Van Der Merwe, M. and Van Vuren, J.H.J. (1997) Bioaccumulation of Selected Metals in African Sharptooth Catfish, Clarias gariepinus from the Lower Olifants River, Mpumalanga, South Africa. Koedoe, 40, 77-90.http://dx.doi.org/10.4102/koedoe.v40i1.265

6. Edward, J.B., Idowu, E.O., Oso, O.R. and Ibidapo, O.R. (2013) Determination of Heavy Metals in Fish Samples, Sediment and Water from Odo-Ayo River in Ado-Ekiti, Ekiti State, Nigeria. International Journal of Environmental Monitoring and Analysis, 1, 27-33.http://dx.doi.org/10.11648/j.ijema.20130101.14

7. Wepener, V. (1997) Metal Ecotoxicology of the Olifants River in the Kruger National Park and the Effect Thereof of Fish Haematology. Ph.D. Dissertation, Rand Afrikaans University, South Africa.

8. Battacharya, A. and Battacharya, S. (2007) Induction of Oxidative Stress by Arsenic in Clarias batrachus: Involvement of Peroxisomes. Ecotoxicology and Environmental Safety, 66, 178-187.

9. Hogstrand, C. and Haux, C. (1991) Mini-Review: Binding and Detoxification of Heavy Metals in Lower Vertebrates with Reference to Metallothionein. Comprehensive Biochemical Physiology, 100C, 383-390.

10. Ventura-Lima, J., Bogo, M.R. and Monserrat, J.M. (2011) Arsenic Toxicity in Mammals and Aquatic Animals: A Comparative Biochemical Approach. Ecotoxicology and Environmental Safety, 74, 211-218. http://dx.doi.org/10.1016/j.ecoenv.2010.11.002

11. Osman, A., Wuertz, S., Mekkawyi, I., Exner, H. and Kirschbaum, F. (2007) Lead Induced Malformations in Embryos of the African Catfish Clarias gariepinus (Burchell, 1822). Environmental Toxicology, 22, 375-389. http://dx.doi.org/10.1002/tox.20272

12. Magadza, C.H.D. (2003) Lake Chivero: A Management Case Study. Lakes & Reservoirs: Research & Management, 8, 69-81. http://dx.doi.org/10.1046/j.1320-5331.2003.00214.x

13. Regina, N.M.M. (2012) Biological Monitoring and Pollution Assessment of the Mukuvisi River, Harare, Zimbabwe. Lakes & Reservoirs: Research & Management, 17, 73-80.http://dx.doi.org/10.1111/j.1440-1770.2012.00497.x

14. Phiri, C. (2000) An Assessment of the Health of Two Rivers within Harare, Zimbabwe, on the Basis of Macroinvertebrate Community Structure and Selected Physicochemical Variables. African Journal of Aquatic Science, 25, 134-145.http://dx.doi.org/10.2989/160859100780177677

15. Mathuthu, A.S., Zaranyika, F.M. and Jonnalagadda, S.B. (1993) Monitoring of Water Quality in Upper Mukuvisi River in Harare, Zimbabwe. Environment International, 19, 51-61. http://dx.doi.org/10.1016/0160-4120(93)90006-4

16. Zaranyika, M.F. (1996) Sources of Levels of Pollution along Mukuvisi River: A Review. In: Moyo, N.A.G., Ed., Lake Chivero: A Polluted Lake, University of Zimbabwe Publications, Harare, 35-42.

17. Machena, C. (1997) The Pollution and Self-Purification Capacity of the Mukuvisi River. In: Moyo, N.A.G., Ed., Lake Chivero: A Polluted Lake, University of Zimbabwe Publishers, Harare, 75-91.

18. Moyo, N.A.G. and Worster, K. (1997) The Effects of Organic Pollution on the Mukuvisi River, Harare, Zimbabwe. In: Moyo, N.A.G., Ed., Lake Chivero: A Polluted Lake, University of Zimbabwe Publishers, Harare, 53-63.

19. Madanire-Moyo, G. and Barson, M. (2010) Diversity of Metazoan Parasites of the African Catfish Clarias gariepinus (Burchell, 1822) as Indicators of Pollution in a Subtropical African River System. Journal of Helminthology, 84, 216- 227.http://dx.doi.org/10.1017/S0022149X09990563

20. Nhiwatiwa, T., Barson, M., Harrison, A.P., Utete, B. and Cooper, R.G. (2011) Metal Concentrations in Water, Sediment and Sharptooth Catfish Clarias gariepinus from Three Peri-Urban Rivers in the Upper Manyame Catchment, Zimbabwe. African Journal of Aquatic Science, 36, 243-252. http://dx.doi.org/10.29 89/16085914.2011.636906

21. Faber, L., Gilcreas, F.W., Edwards, G.P. and Taras, M.J. (1960) Standard Methods for the Examination of Water and Wastewater: Including Bottom Sediments and Sludges. 11th Edition, American Public Health Association, Inc., New York.

22. Greenberg, A.E., Connors, J.J. and Jenkin, D. (1980) Standard Methods for the Examination of Water and Wastewater. 15th Edition, American Public Health Association, Washington DC.

23. Hammer, O., Harper, D.A.T. and Ryan, P.D. (2012) PAST—Palaeontological Statistics. Version 1.90. http://folk.uio.no/ohammer/past

24. Weiner, J.G. and Giesy Jr., J.P. (1979) Concentrations of Cd, Cu, MN, Pb & Zn in Fishes in a Highly Organic Softwater Pond. Journal of the Fisheries Research Board of Canada, 36, 270-279. http://dx.doi.org/10.1139/f79-042

25. TerBraak, C.J.F. and Šmilauer, P. (2002) CANOCO Reference Manual and CanoDraw for Windows User's Guide: Software for Community Ordination. Version 4.5, Microcomputer Power, Ithaca, NY.

26. Milenkovic, N.M., Damjanovic, M. and Ristic, M. (2005) Study of Heavy Metal Pollution in Sediments from the Iron Gate (Danube, River), Serbia and Montenegro. Polish Journal of Environmental Studies, 14, 781-787.

27. Ali, A., Ahmadou, D., Adji Mohamadou, B., Saidou, C. and Tenin, D. (2010) Determination of Minerals and Heavy Metals in Water, Sediments and Three Fish Species (Tilapia nilotica, Silurus glanis and Arius parkii) from Lagdo Lake Cameroun. Journal of Fisheries International, 5, 54-57.

28. Alexopoulous, E., McCrohan, C.R., Powell, J.J., Juqidaohsingh, R. and White, K.N. (2003) Bioavailability and Toxicity of Freshly Neutralized Aluminium to the Freshwater Crayfish (Pacifastacus leniusculus). Archives of Environmental Contamination and Toxicology, 45, 509-514. http://dx.doi.org/10.1007/s00244-003-0228-9

29. Canli, M. and Kalay, M. (1998) Levels of Heavy Metals (Cd, Pb, Cu, Cr and Ni) in Tissue of Cyprinus carpio, Barbus capito and Chondrostoma regium from the Seyhan River, Turkey. Turkish Journal of Zoology, 22, 149-157.

30. Trenfield, M.A., Morkich, S.J., Nq, J.C., Noller, B. and van Dam, R.A. (2012) Dissolved Organic Carbon Reduces Toxicity of Aluminium to Three Tropical Freshwater Organisms. Environmental Toxicology and Chemistry, 31, 427- 436.http://dx.doi.org/10.1002/etc.1704

31. Poleo, A.B., Ostbye, K., Oxnerard, S.A., Andersen, R.A., Heibo, E. and Vollestad, L.A. (1997) Toxicity of Acid Aluminium Rich Water to Seven Freshwater Species: A Comparative Laboratory Study. Environmental Pollution, 96, 129- 139.http://dx.doi.org/10.1016/S0269-7491(97)00033-X

32. Mohamed, Z.A. (2001) Removal of Cadmium and Manganese by a Non-Toxic Strain of the Freshwater Cyanobacteria Gloethece magna. Water Research, 35, 4405-4409.http://dx.doi.org/10.1016/S0043-1354(01)00160-9

33. Lasier, P.J., Winger, P.V. and Bogencider, K. (2000) Toxicity of Manganese to Ceriodaphnia dubia and Hyalella azteca. Archives of Environmental Contamination and Toxicology, 38, 298-304. http://dx.doi.org/10.1007/s002449910039

34. Karel, A.C., Deschamphelare, J., Koene, D.G. and Jansen, C.R. (2008) Reduction of Growth and Haemolymph Ca Levels in the Freshwater Snail Lymnaea stagnalis Chronically Exposed to Cobalt. Ecotoxicology and Environmental Safety, 71, 65-70. http://dx.doi.org/10.1016/j.ecoenv.2007.07.004

35. WHO (2006) Guidelines for Drinking Water Quality. Drinking Water Quality Control in Small Community Supplies, Geneva, 3, 121.

Chapter 7

Phosphate-Mediated Remediation of Metals and Radionuclides

Robert J. Martinez, Melanie J. Beazley, and
Patricia A. Sobecky

Department of Biological Sciences, University of Alabama, 300
Hackberry Lane, Tuscaloosa, AL 35487, USA

ABSTRACT

Worldwide industrialization activities create vast amounts of organic
and inorganic waste streams that frequently result in significant soil
and groundwater contamination. Metals and radionuclides are of
particular concern due to their mobility and long-term persistence
in aquatic and terrestrial environments. As the global population
increases, the demand for safe, contaminant-free soil and groundwater
will increase as will the need for effective and inexpensive remediation
strategies. Remediation strategies that include physical and chemical

methods (i.e., abiotic) or biological activities have been shown to impede the migration of radionuclide and metal contaminants within soil and groundwater. However, abiotic remediation methods are often too costly owing to the quantities and volumes of soils and/or groundwater requiring treatment. The in situ sequestration of metals and radionuclides mediated by biological activities associated with microbial phosphorus metabolism is a promising and less costly addition to our existing remediation methods. This review highlights the current strategies for abiotic and microbial phosphate-mediated techniques for uranium and metal remediation.

INTRODUCTION

The global population is predicted to reach 10 billion by the year 2100 [1]. To support the demand for increased food production and access to fresh water, human societies will be forced to employ less desirable (i.e., lower quality) soil and groundwater resources for crop production and drinking water [1–3]. Human activities associated with 20th century industrial-scale production of electrical components, fabrics, fertilizers, inks and dyes, mining, metal production, paints, paper products, pesticides, pharmaceuticals, rubber, and plastics contribute to the degradation of surface and subsurface sediments and water quality as evidenced by the production of more than 1 million metric tons of metal waste per year [4, 5]. Governmental activities have also contributed to the contamination of soils and groundwater throughout the United States, where the legacy of nuclear weapons research and development has resulted in the contamination of estimated 75 million cubic meters of sediment and more than 1.8 billion cubic meters of groundwater [6]. The devastating 2011 earthquake of the coast of Japan and the subsequent tsunami that destroyed three nuclear reactors at the Fukushima Daiichi Nuclear Power Plant complex highlight a more recent challenge to remediation and disposal of large quantities of nuclear fuel materials including radionuclides. The scope of the remediation challenge is considerable, as deleterious effects to human health, surrounding environment, and food supply are well known due to metal and radionuclide exposure and ingestion [7–10]. Moreover, the presence of these contaminants negatively impacts ecosystem sustainability and contribute to the loss of biodiversity [11].

Lead (Pb), cadmium (Cd), and zinc (Zn) represent a subset of the most frequently reported metal contaminants in sediments and groundwater. These metals, many of which cooccur at the same site, are listed on the US EPA National Priorities List and are detected at many US EPA Superfund sites [12]. Uranium (U) waste, resulting from U.S. nuclear weapons production, is found in soils and groundwater at the Department of Energy (DOE) facilities [13] and in coal and phosphate mining/processing waste sites [14–19]. The fate and transport of metals and radionuclides in the environment are controlled by geochemical factors, including pH, adsorption, reduction/oxidation, and precipitation reactions. In natural environments, pH is one of the primary controlling variables for metal and radionuclide speciation [20]. Below pH ~5, most metals and radionuclides tend to exist primarily as free divalent cations and as solid oxyhydroxides, carbonates, and oxides above pH 7 (Figures 1(a) and 1(b)). In the presence of phosphate, precipitation reactions control metal speciation through the formation of highly insoluble metal- and radionuclide-phosphate minerals that are stable over a wide pH range (Figures 2(a) and 2(b)). Though less studied to date than other methods, remediation approaches that promote phosphate immobilization of metals and radionuclides represent viable strategies for long-term in situ sequestration.

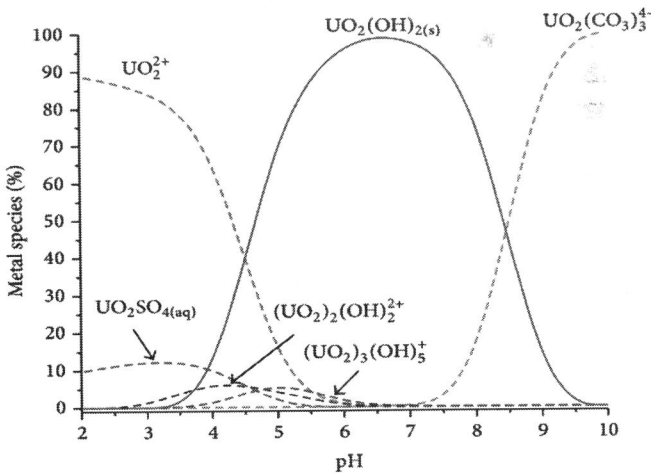

(a)

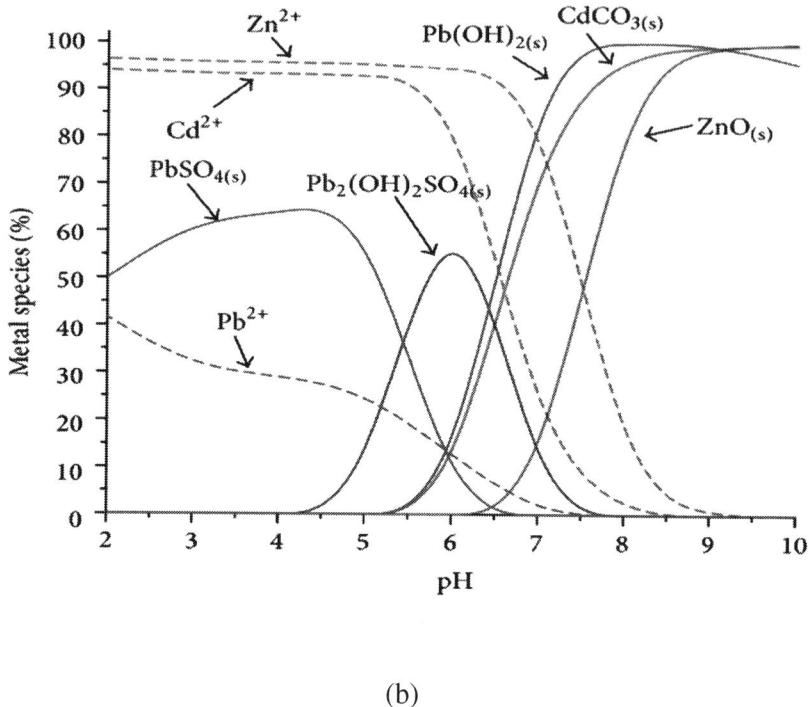

(b)

Figure 1: Influence of changing pH on chemical speciation of (a) U(VI) and (b) metal mixture composed of Cd(II), Pb(II), and Zn(II). Calculated using MINEQL + v. 4.5 and the Nuclear Energy Agency's updated thermodynamic database for uranium [106, 107] as a function of pH in synthetic ground-water (SGW). The open system model at 22°C calculated aqueous (dotted lines) and solid phases (solid lines) at equilibrium using the concentrations of ions present in SGW, $UO_2^{2+}{}_{(aq)} = 500\ \mu M$, $Cd^{2+}{}_{(aq)} = 500\ \mu M$, $Pb^{2+}{}_{(aq)} = 500\ \mu M$, $Zn^{2+}{}_{(aq)} = 500\ \mu M$, and $P_{CO2} = 10^{-3.5}$ atm. Chemical speciation calculations utilized SGW that consists of 2 μM FeSO$_4$, 5 μM MnCl$_2$, 8 μM Na$_2$MoO$_4$, 0.8 mM MgSO$_4$, 7.5 mM NaNO$_3$, 0.4 mM KCl, 7.5 mM KNO$_3$, and 0.2 mM Ca (NO$_3$)$_2$ [57].

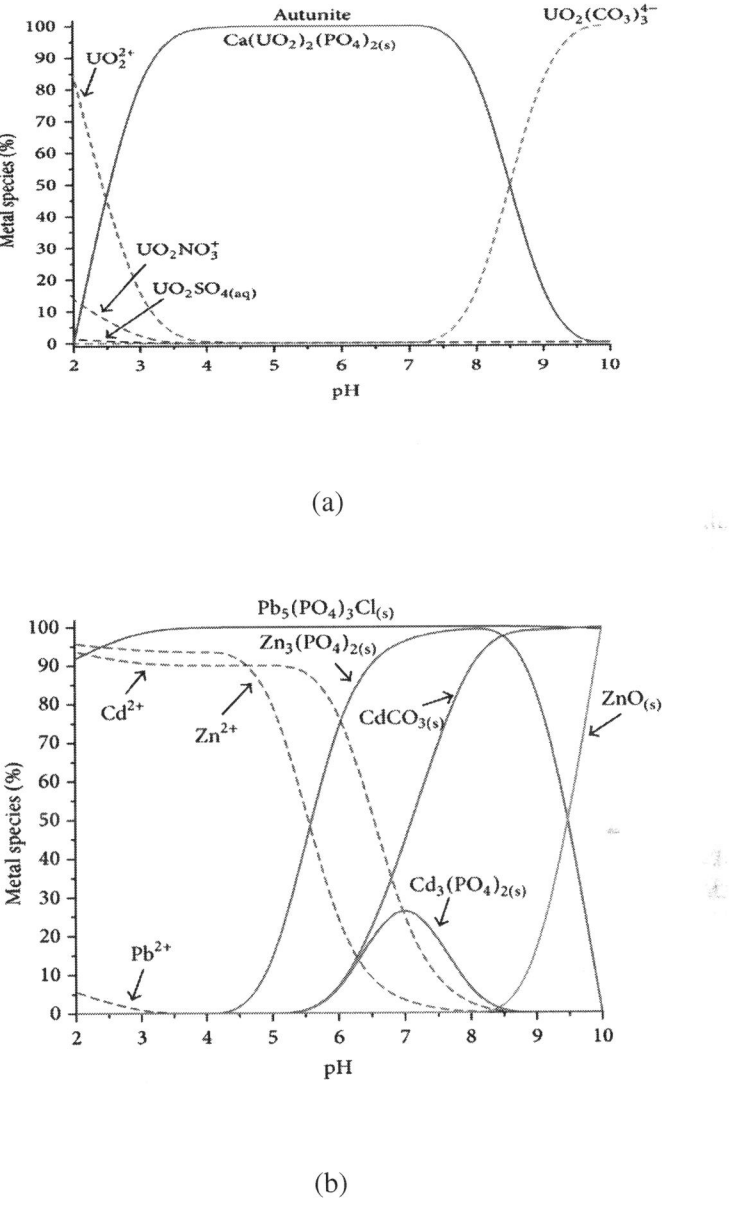

(a)

(b)

Figure 2: Influence of soluble phosphate and changing pH on chemical speciation of (a) U(VI) and (b) metal mixture composed of Cd(II), Pb(II), and Zn(II), Calculated using MINEQL + v. 4.5 and the Nuclear Energy Agency's updated thermodynamic database for uranium [106, 107]

as a function of pH in synthetic groundwater (SGW). The open system model at 22°C calculated aqueous (dotted lines) and solid phases (solid lines) at equilibrium using the concentrations of ions present in SGW, $UO_2^{2+}{}_{(aq)}$= 500 μM, $Cd^{2+}{}_{(aq)}$= 500 μM, $Pb^{2+}{}_{(aq)}$=500 μM, $Zn^{2+}{}_{(aq)}$= 500 μM, and P_{CO2}= $10^{-3.5}$ atm. Chemical speciation calculations utilized SGW that consists of 2 μM $FeSO_4$, 5 μM $MnCl_2$, 8 μM Na_2MoO_4, 0.8 mM $MgSO_4$, 7.5 mM $NaNO_3$, 0.4 mM KCl, 7.5 mM KNO_3, and 0.2 mM Ca $(NO_3)_2$ [57].

Strategies for the remediation of metal and radionuclide-contaminated soils and groundwater include physical and chemical (i.e., abiotic) and biologically mediated methods. Chemical and physical methods, including excavation and soil capping [21, 22], pump and treat technologies [23, 24], mineral adsorption [25–27], mineral precipitation [28, 29], complexation [30–32], adsorption to permeable zero-valent iron and hydroxyapatite reactive barriers [33–35], cement solidification [36, 37], and vitrification [38, 39], have all demonstrated efficacy in mitigating contaminant transport in situ. Remediation methods that depend upon chemical transformations for in situ sequestration of contaminants must first consider local geochemical parameters that include local geology, concentrations of soluble anions and cations, pH, and redox state. The influence of pH on a contaminated groundwater system (Figures 1(a) and 1(b)) highlights the importance of such considerations in order to predict contaminants speciation and bioavailability [40–44].

In addition to the chemically mediated methods for contaminant immobilization, bioremediation has demonstrated great promise as an additional strategy to promote in situ sequestration of contaminants [11,45–48] by harnessing the metabolisms of plants and microorganisms to detoxify and/or affect the in situ mobility of a given contaminant. The use of prokaryotic and eukaryotic microorganisms has proven effective for metal and radionuclide remediation through the processes of biosorption, bioaccumulation, bioreduction, and biomineralization [49–61]. Recently, it has been shown that microorganisms, which can hydrolyze organophosphate compounds with a concomitant increase in the liberation of extracellular orthophosphate to the surrounding environment, represent a unique approach to promote subsurface in situ phosphate biomineralization of metals and radionuclides. The stability of phosphate minerals over a broad pH range (Figures 2(a) and 2(b)) provides an ideal insoluble phase for long-term contaminant sequestration within subsurface environments that experience changing local geochemistry (e.g., Eh, pH, oxidants).

The application of various phosphate compounds (e.g., orthophosphate solutions, soluble polyphosphates, and organophosphates) to immobilize contaminants (e.g., Cd, Pb, U, and Zn) in laboratory as well as field experiments detected contaminant sequestration via a combination of microbial-mediated mechanisms and abiotic reactions. The focus of this review will be on the chemical (abiotic) and microbial mechanisms promoting phosphate-mediated immobilization of uranium and cooccurring metals (Figure 3) within subsurface environments as well as highlighting the unique challenges that dynamic geochemical conditions have on long-term in situ sequestration strategies.

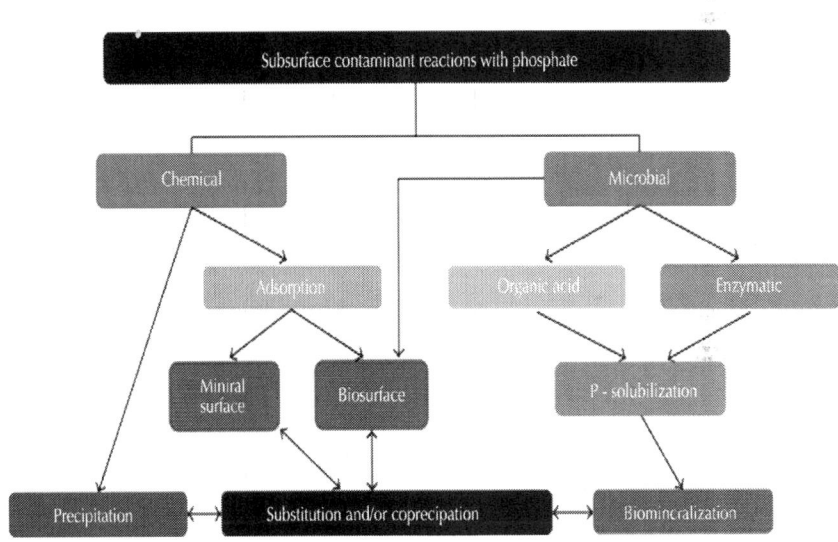

Figure 3: Flow-chart summary of chemical- and microbial-mediated reactions with phosphate that promote immobilization of metal and radionuclide contaminants within subsurface environments.

CHEMICAL APPROACHES TO PHOSPHATE IMMOBILIZATION OF METALS AND RADIONUCLIDES

The chemistry of phosphate is such that it facilitates reactions with over 30 elements and has resulted in the identification of 300 unique phosphate minerals [62]. On Earth, 95% of all phosphorus (P) is present within apatite minerals $[Ca_5(PO_4)_3(F,OH,Cl)]$ [63, 64] and the biogeochemical interactions of this mineral not only play a role in global P cycling but also influence the mobility of soluble metals [65]. The use of orthophosphate solutions has been shown to immobilize contaminants (e.g., Cd, Pb, U, and Zn) via mineral precipitation over a broad pH range (Figures 2(a) and 2(b)) [66–69]. Additionally, phosphate minerals promote cocontaminant sequestration via mineral substitution reactions, coprecipitation reactions, adsorption, and ion-exchange reactions and provide a source of orthophosphate for dissolution-precipitation reactions [34, 70–76]. Various soluble and insoluble P compounds (e.g., orthophosphate solutions, soluble polyphosphate, phosphatidic clays, apatite reactive barriers, and vivianite) have demonstrated success in metal and radionuclide sequestration [34,68, 77–85].

Soluble Phosphate Amendments

Soluble P, in the forms of phosphoric acid, phytic acid, and tripolyphosphate, has been examined for the sequestration of Cd, Cu, Pb, U, and Zn in contaminated environments [78, 82, 84, 86–88]. Phosphoric acid and inorganic phosphate salts are examples of simple forms of reactive orthophosphate effective in forming stable minerals in surface soils (depth < 12.5 cm) contaminated with Cd, Cu, Pb, and Zn [87–89]. The use of orthophosphate solutions within surface soils and deep subsurface environments requires strategies that prevent (1) rapid precipitation of contaminants and other sediment components (e.g., Al, Ca, Fe, Mg, and Mn) that affects hydraulic conductivity, (2) leaching of metals (e.g., As, Se, and W), and (3) ecosystem eutrophication caused by excess P runoff [82, 89–91].

Compounds such as phytic acid (i.e., the acid form of inositol-6-phosphate) and polyphosphate have been examined for their ability to (1) act as sites for ion exchange, (2) facilitate the slow delivery of orthophosphate, and (3) act as chelating agents that minimize the bioavailability of cations within contaminated environments [92–95]. The use of these compounds will ideally minimize undesirable phosphate-sediment interactions at the site of delivery, allow for greater mobility beyond the point of injection within subsurface sediments, and promote greater sequestration of metals and/or radionuclides when compared to injections of orthophosphate-rich solutions.

Laboratory studies examining the immobilization of various cations (i.e., Ba, Co, Mn, Ni, Pb, U, and Zn) within contaminated sediments have shown phytic acid to be effective in reducing soluble contaminant concentrations. Specifically, concentrations of U(VI) were reduced from $2,242\,g\,kg^{-1}$ to $76\,g\,kg^{-1}$ and of Ni from $58\,mg\,kg^{-1}$ to $9.6\,mg\,kg^{-1}$ in sediment batch incubations treated with calcium phytate (i.e., the salt of phytic acid) [92, 93]. However, the solubility of these metal-phytate complexes is highly dependent upon ionic strength, pH, ligand conformation, and metal-to-ligand ratio [96], which limits the efficacy of phytic acid for metal and radionuclide immobilization in dynamic geochemical conditions. More recent studies of phytic acid, sodium monophosphate, and sodium tripolyphosphate (TPP) interactions with U(VI) conducted in column experiments investigated sediments with basic porewater pH and geochemical conditions representative of metal- and radionuclide-contaminated environments in the western U.S. [82, 84]. In column studies, without soluble U, cation interactions with phytic acid and sodium monophosphate resulted in a 30% decrease in hydraulic conductivity [82]. Conversely, TPP amendments did not affect hydraulic conductivity and allowed for a slow release of orthophosphate within the sediment column [82]. Within columns containing over $100\,mg\,L^{-1}$ soluble U, TPP amendments were capable of promoting rapid precipitation of U, thereby decreasing U concentrations below U.S. EPA drinking water standards ($30\,\mu g\,L^{-1}$) [84]. The use of TPP, rather than phytic acid or sodium monophosphate, demonstrates great promise within neutral-to-alkaline sediments as an agent for metal and radionuclide sequestration.

Solid Reactive Phosphate Amendments

An alternative to introducing soluble orthophosphate into contaminated sediment and groundwater is the application of apatite $[Ca_5(PO_4)_3(F,OH,Cl)]$ minerals (e.g., bone apatite, synthetic apatite minerals, and rock phosphate) as subsurface reactive barriers [97–102]. Numerous studies have demonstrated that in situ soluble metal and radionuclide concentrations can be greatly reduced by surface interactions, dissolution-precipitation reactions, and ion-exchange with apatite minerals [34, 70, 73, 103, 104]. Apatite minerals have been effective in decreasing soluble concentrations (i.e., a 1,000-fold reduction) of contaminants that include Cd, Co, Cu, Hg, Mn, Ni, Pb, Sb, Th, and U [73, 79, 105]. Apatite reactive barriers have been effective in decreasing soluble metal and radionuclide concentrations in situ; however, the reversibility of cation adsorption and possible changes to hydraulic conductivity must be considered with this approach [34, 80, 81, 100]. Thus, the use of apatite requires continuous monitoring of the barrier effluent under dynamic geochemical conditions (i.e., changing contaminant concentration and pH) to maintain optimal contaminant sequestration and to calculate barrier lifetime [80, 100].

Recent studies have demonstrated promising results in the use of phosphate mineral nanoparticles for the remediation of Cu, Cd, and Pb contaminants [83, 108, 109]. The use of hydroxyapatite and vivianite $[Fe_3(PO_4)_2]$ nanoparticles in the size range of 3–10 nm provides properties of both liquid and solid phosphate amendments by facilitating liquid injection of particles that have a high surface area. Additionally, the use of nanoparticles offers an approach that maximizes contaminant immobilization by minimizing (1) orthophosphate loss to runoff, (2) precipitation reactions at the injection site, and (3) leaching of oxyanions. The unique application of phosphate mineral phases in the nanoparticle size range not only offers an effective delivery system to promote contaminant adsorption, precipitation, and/or ion-exchange reactions, but also provides a source of P to deep subsurface microbial communities capable of catalyzing reactions that further contribute to contaminant sequestration. Due to the lack of studies that examine nanoparticle fate and transport as well as the difficulties in monitoring this class of material in the environment, the potential hazards posed to prokaryotes, plants, and animals must be evaluated prior to their implementation in any remediation strategies [110–112].

BIOLOGICAL APPROACHES TO PHOSPHATE-MEDIATED IMMOBILIZATION OF METALS AND RADIONUCLIDES

Phosphorus in the form of inorganic phosphate (P_i) is essential for cellular energy conservation and proper structure/function of cellular macromolecules (i.e., nucleic acids, proteins, sugars, and lipids) in all living organisms [113]. Due to the essential requirement for P, the scarcity of this nutrient within terrestrial environments (i.e., average of 1 µM [114]) requires plants, fungi, and bacteria to employ P-scavenging strategies. Bacterial acquisition of P can occur through secretion of organic acids that solubilize phosphate minerals via expression of organophosphate hydrolases [115–124]. Conversely, microbial stress (i.e., low nutrients, low pH, and oxidizing agents) results in storage of P as intracellular polyphosphate [125–127]. The following sections will specifically examine microbial polyphosphate metabolism and organophosphate hydrolase activity as they relate to metal and radionuclide immobilization.

Polyphosphate-Mediated Bioaccumulation

The biologically mediated polymerization of orthophosphate molecules as intracellular polyphosphate granules is among the most ancestral metabolic functions present in all domains of life [125]. In bacteria, this polymer has been shown to function as an energy source, P reserve, metal chelator, and a regulator of stress and development [125, 127]. Polyphosphate metabolism in E. coli and other bacterial species is driven by the genes polyphosphate kinase (ppk) and exopolyphosphatase (ppx) that catalyze intracellular inorganic phosphate polymerization and hydrolysis of polyphosphate granules, respectively [125]. Phenotypic analyses of ppk knock-out mutations have expanded the known functions of polyphosphates that contribute to virulence, motility, biofilm formation, and sensitivity to stressors (e.g., oxidative, heat, osmotic, pH, and nutrient) [125–127]. Within metal and radionuclide contaminated environments, low nutrient availability, high concentrations of oxidizing agents, and extremes in pH typify chemical stressors that pose challenges to microbial metabolism. In response to the local geochemistry of contaminated environments, production of intracellular polyphosphate provides microorganisms with a means to sequester toxic ions within the cell cytosol as well as the regulation of gene(s) expressed in response to cellular stress (e.g., DNA repair, RNA polymerase sigma factor, and pH extremes) [55, 127–131].

The reactivity of cytosolic polyphosphates has been shown to facilitate intracellular sequestration of Cd, Cu, Hg, Pb, U, and Zn in genetically engineered bacterial strains as well as naturally occurring archaeal and bacterial strains [52, 55, 132–136]. Electron microscopy analyses of cells exposed to these elements demonstrated intracellular localization with phosphate-rich granules, suggesting that contaminant sequestration may be achieved by polyphosphates and may protect sensitive cytosolic molecules from oxidative damage. In addition to polyphosphate chelation of metals and radionuclides, an engineered Pseudomonas aeruginosa strain overexpressing the ppk gene was shown to enhance intracellular phosphate concentrations when compared to the wild-type strain [133]. Upon nutrient starvation, polyphosphate depolymerization and efflux of phosphate into the media containing U(VI) promoted the removal of 80% of the soluble U(VI) as a uranyl phosphate precipitate. Thus, polyphosphate metabolism that promotes intracellular or extracellular sequestration of metals and radionuclides represents a remediation approach that harnesses the physiologies of extant microorganisms within contaminated environments.

Phosphatase-Mediated Biomineralization

Redox-independent biomineralization of metals and radionuclides, although not as extensively studied as reductive precipitation to date, could provide a complementary approach to the existing remediation strategies. Biogenically derived phosphate minerals result from the activities of microbial phosphatases, enzymes that are essential for microbial acquisition of C and P, regulation of cellular metabolism, and signal transduction [137–139]. Phosphatase enzymes are classified by pH optima (acid or alkaline), molecular weight, and those that hydrolyze only phosphorylated serine or threonine residues [140]. Acid and alkaline phosphatases (either cytoplasmic or periplasmic) are fundamentally required for microbial nutrient acquisition [116, 119, 141–143].

Early work examining bacteria capable of phosphate-mediated biomineralization of metals and radionuclides focused on the acid phosphatase activity of Serratia sp. NCIMB 40259 (formerly Citrobacter species) [50, 144,145]. Several studies that characterized Class A NSAP of Serratia sp. NCIMB 40259 reported that microorganisms with similar phosphatase activities could be exploited to promote mineralization of metals [50, 144–146]. Additionally, biofilms of Serratia sp. NCIMB 40259 that promoted the precipitation of $H_2(UO_2)_2(PO_4)_2$ (chernikovite) further removed 85% and 97% of cooccurring ^{60}Co and ^{137}Cs, respectively, via substitution of H^+ within chernikovite [147].

Prior to investigations into a possible bioremediation role for nonspecific acid phosphatases (NSAP), this class of enzymes had been examined for their role in microbial physiology and contributions to virulence [116, 148]. Although the exact physiological roles for these enzymes have yet to be discerned, the biochemical properties of NSAPs have been well studied. NSAPs are divided into three unique classes (Classes A, B, and C) based on catalytic domain motifs, comprised of low molecular weight monomeric subunits ranging from 25 to 30 kDa, and have a pH optima ranging from 5.0 to 6.5 and catalytic activities for a broad range of phosphomonoester substrates [116]. NSAPs can be localized to the outer membrane, periplasmic space and/or are secreted into the extracellular environment. As the physiological properties of bacterial NSAPs facilitate their activities at low pH conditions typical of many mixed waste sites, their contributions to in situ metal detoxification and metal immobilization activities may be greatly underappreciated.

Within mixed waste environments, the in situ activities of microbial phosphatases likely contribute to localized cation precipitation in addition to nutrient acquisition. Studies examining phosphatase activities (e.g., alkaline or acid) of genetically engineered and naturally occurring strains of Gram-positive and Gram-negative bacteria were shown to promote U immobilization (>90% precipitation of soluble U) via precipitation and coprecipitation reactions [50, 57, 60, 74, 149–153]. Interestingly, soil bacterial isolates demonstrated constitutive phosphatase activities that liberated comparable, if not greater concentrations, of reactive phosphate when compared to the phosphatase activities of genetically modified strains [57, 149]. The occurrence of naturally occurring bacteria with constitutive phosphate-liberating NSAP phenotypes isolated from radionuclide and metal contaminated subsurface soils supports the role of acid phosphatases in promoting microbial adaptation to in situ metal stresses.

To further highlight such adaptation, recent investigations of terrestrial and marine bacterial isolates belonging to the genera Aeromonas, Bacillus, Myxococcus, Pantoea, Pseudomonas, Rahnella, and Vibrio demonstrated that Cr, Pb, and U were removed from solution as phosphate minerals under both oxic and anoxic growth conditions [57, 60, 74, 152, 154–158]. Our recent work further examined lead and uranium precipitates produced by Rahnella sp. Y9602, using X-ray diffraction (XRD), variable pressure scanning electron microscopy/ energy dispersive X-ray spectroscopy (VP-SEM/EDX), and extended X-ray absorption fine structure (EXAFS). Hydrated lead precipitates that accumulated on the cell surface of Rahnella sp. Y9602 were identified as lead hydroxyapatite, $Pb_{10}(PO_4)_6(OH)_2$ by XRD analysis (data unpublished). VP-SEM/EDX elemental mapping demonstrated U and P localization in precipitates generated by Rahnella sp. Y9602 grown in minimal media containing soluble U(VI). Subsequent synchrotron-based XRD and EXAFS analyses of these uranium phosphate precipitates identified the mineral as chernikovite $[H_2(UO_2)_2(PO_4)_2]$ [60].

CHALLENGES FOR IN SITU IMMOBILIZATION OF METALS AND RADIONUCLIDES

The goal of metal and/or radionuclide remediation strategies that focus on in situ sequestration is to generate an insoluble precipitate that will be immobilized, sequestered, and stable within a given environment for effective long-term stewardship. The difficulty in maintaining precipitates in the solid phase arises from the dynamic nature of biogeochemical processes in the environment.

Approaches that employ reductive precipitation of metals and radionuclides through the use of zero-valent iron reactive barriers, mineral phase interactions, and microbial reduction activities face challenges for long-term sequestration that arise from oxidant (e.g., oxygen, nitrate, manganese, iron oxides, or humic substances) interactions that remobilize precipitated reduced contaminants to their oxidized valence states, thereby increasing their solubility [19, 45, 159–168]. Limitations to microbial reduction of metals and radionuclide also arise from low pH environments that must be neutralized to support growth of sulfate- and metal-reducing communities [163, 169, 170]. Additionally, microbial reduction in mixed waste environments must be carefully considered so as to minimize contaminant migration of elements such as As and Pu that demonstrate greater mobility in their reduced valence states [171, 172].

In contrast, the geochemical stability of insoluble metal- and radionuclide-phosphates allows for in situsequestration within environments that undergo dynamic changes in redox conditions and pH. Such changes, in local geochemistry, however, do not support stable sequestration of metals and radionuclides in their reduced forms. Within uranium contaminated environments, phosphate mineralization of soluble U(VI) produces a wide array of uranyl phosphate minerals [173]. The formation of these minerals has been implicated in the control of U mobility in U.S. DOE contaminated sediments [174, 175], although the microbial contribution to in situ uranyl phosphate formation in these DOE sites has yet to be determined. With recent studies identifying autunite- and hydroxyapatite-precipitating capabilities of Aeromonas, Bacillus,Pantoea, Pseudomonas, and Rahnella spp. in both oxic and anoxic growth conditions, the synergistic properties of these minerals (i.e., ion-exchange reactions that sequester cooccurring metals) highlight an important role in not only stabilizing U contamination but also cooccurring metals [57, 60, 74, 157]. Furthermore, contaminated sites that are characterized by acidic to circumneutral porewater pH represent environments that can support stable mineral formation (Figures 2(a) and 2(b)), provided that carbonates are not present in significant concentrations (i.e. $P_{CO2} < 10^{-3.5}$ atm) [176, 177]. Interestingly, investigations of microbial reduction of Cr, Np, Pu, and U have been shown to support subsequent phosphate precipitation reactions via thermodynamic modeling, chromatographic separation of actinides based on valence state, and X-ray analytical methods [154, 155, 172, 178, 179]. Unlike U, that is capable of forming phosphate minerals in both hexavalent and tetravalent states [50, 179], the reduction of Cr, Np, and Pu is initially required for these contaminants to participate in phosphate precipitation reactions [154, 155, 172, 178]. To date, only pure culture or coculture studies have identified such coupled microbial interactions that offer an additional approach to control contaminant toxicity and mobility that are perpetuated by valence state cycling. Due to the limited number of studies, future work is required to further understand protocooperative interactions between extant subsurface metal reducing and phosphate solubilizing microbial communities that promote contaminant sequestration. Additionally, further examination of reduced valence state contaminants precipitated as phosphate minerals are required to understand the influence of changing geochemical parameters (e.g., Eh, pH, and oxidants) that can affect solubility of the in situ immobilized contaminants.

Overall, these studies highlight the need to consider contaminant physicochemical properties, redox state of the environment, pH, presence of complexing ligands, and the metabolic properties of the extant microbial community when developing an in situ sequestration strategy.

SUMMARY, CHALLENGES, AND FUTURE DIRECTIONS

Prior to anthropogenic releases of metals and radionuclides into the environment, these elements had been (and continue to be) discharged into the environment through volcanic activity, hydrothermal vent sources, and the dissolution of metal-bearing minerals [180]. Prokaryotic and eukaryotic organisms play key roles in geochemical cycling through metabolic processes that scavenge, mobilize, and precipitate these elements in terrestrial, aquatic, and atmospheric environments [180–182]. The evolution of prokaryotic and eukaryotic organisms that influence the solubility of metal- and radionuclide-bearing minerals while tolerating changing concentrations of these elements has given rise to microbial metabolic diversity beneficial to passive and active environmental remediation efforts.

Passive remediation approaches such as monitored natural attenuation (MNA) and in situ sequestration of metals and radionuclides offer economical alternatives that minimize human exposure to contaminants. Regardless of the type of implemented remediation approach, successful in situ sequestration of metals and radionuclides requires the contaminant(s) remain immobilized as an insoluble species. Implementation of MNA of metals and radionuclides relies on natural physical, chemical, and biological processes to remediate a contaminated environment through the effects of dispersion, dilution, sorption, volatilization, radioactive decay, stabilization, and transformation [183]. Prior to adopting MNA as a sole strategy for metal and radionuclide remediation, the processes by which the contaminant(s) are immobilized must be shown to be irreversible [183]. Unfortunately, this approach relies on limited contaminant plume mobility as well as stable geochemical and hydrological conditions [183–187]. Therefore, contaminated subsurface environments with changing hydrobiogeochemical conditions (typical for most sites) will likely influence speciation chemistry and thus require an alternative strategy.

Active remediation strategies that promote metal- and/or radionuclide-phosphate formation can take advantage of in situ hydrobiogeochemical parameters (Figure 4) that support contaminant sequestration through the formation of (1) low solubility minerals that are unaffected by changes in redox, (2) mineral stability across a wide pH range, and (3) reactive mineral surfaces that can support sequestration of other cooccurring metals through adsorption, substitution, and precipitation reactions [60, 83, 109, 147, 154, 188]. Innovative bioremediation approaches that neutralize low pH groundwater and maintain sufficient phosphate concentrations to complex soluble contaminants can enhance strategies that rely solely on abiotic approaches. By employing a phosphate-mediated pH buffering system, soluble U(VI) can be sequestered as insoluble uranyl hydroxide and uranyl phosphate species [157, 189]. Unlike uraninite (UO_2) mediated sequestration, uranyl phosphate species are not prone to dissolution in oxidizing environments. Thermodynamic modeling and recent anaerobic biomineralization assays utilizing Rahnella sp. Y9602 and U contaminated sediments have shown that uranium phosphate formation can be promoted by microbial hydrolysis of organophosphate substrates under reducing conditions [60, 190].

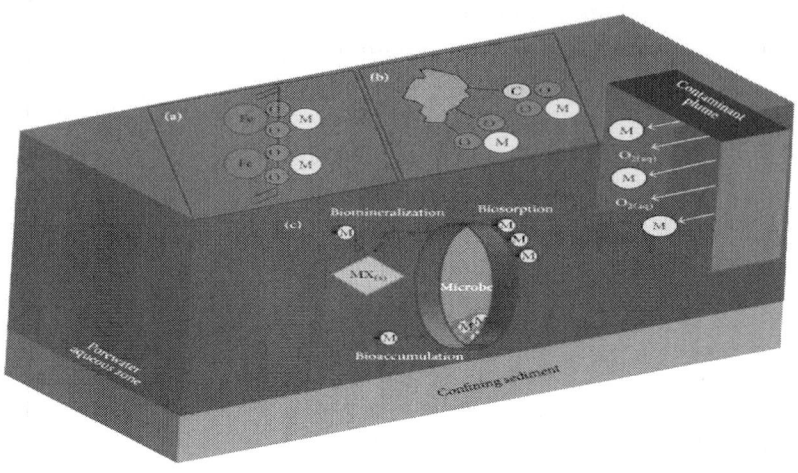

Figure 4: Subsurface hydrobiogeochemical conceptual model. Important chemical and microbial mediated processes that control metal speciation and contaminant transport within subsurface sediments include (a) Fe-hydroxide surface (clays) adsorption; (b) sediment organic matter adsorption; (c) microbial detoxification mechanisms contributing to microbial survival, adaptation, and contaminant immobilization. Microbial metabolic processes that increase local concentrations of anions (X^{n-}; e.g.PO_4^{3-},S^{2-}, and CO^{3-}) promote inorganic contaminant (M^{n+}) mineralization [$MX_{(s)}$].

To date, the majority of metal and radionuclide immobilization studies have focused on bacterial physiology and ecology without considering concomitant archaeal and fungal community responses and/or contributions made by members of these two domains of life. Additional studies that include detailed analyses of archaeal, bacterial, and fungal activities that may be contributing to the mineralization of metals and radionuclides through phosphate-driven mechanisms are needed if we are to have a complete understanding of microbial community responses as they relate to phosphate-mediated bioremediation strategies [122, 135, 191–193]. Interdisciplinary studies that combine analyses of microbial communities and geochemical dynamics are essential to provide a greater understanding of in situ processes that affect contaminant sequestration.

New and emerging methods, such as metagenomics and metaproteomics, which allow for a greater understanding of microbe-metal interactions, many of which were pioneered as a result of academic and U.S. Department of Energy National Laboratory collaborations, have yielded significant insights into subsurface microbial community dynamics and physiological responses within contaminated environments [194–206]. Additionally, synchrotron X-ray techniques (e.g., XRD, XANES, and EXAFS) have become tools for biogeochemical studies that enhance our understanding of in situ contaminant sequestration. Recent studies that incorporate X-ray techniques have elucidated U interactions with sediment surfaces and biomass as well as facilitated mineral identification [60, 153, 179, 207–209]. Combining results from such interdisciplinary studies need to be placed in the context of the whole "in terra" system for long-term field-scale applications [210]. Ultimately, the combined efforts of interdisciplinary research focused on microbial P cycling will support development of predictive models necessary to understand the challenges of long-term contaminant sequestration within geochemically dynamic environments.

ACKNOWLEDGMENT

The authors wish to thank the U.S. Department of Energy, Office of Science (BER), for providing support through U.S. Department of Energy, Grant no. DE-FG02-04ER63906.

REFERENCES

1. NRC, Our Common Journey: A Transition Toward Sustainability, National Academies Press, Washington, DC, USA, 1999.

2. D. Tilman, K. G. Cassman, P. A. Matson, R. Naylor, and S. Polasky, "Agricultural sustainability and intensive production practices," Nature, vol. 418, no. 6898, pp. 671–677, 2002.

3. H. Bouwer, "Integrated water management for the 21st century: problems and solutions," Journal of Irrigation and Drainage Engineering, vol. 128, no. 4, pp. 193–202, 2002.

4. B. L. Morris, A. R. L. Lawrence, P. J. C. Chilton, et al., Groundwater and its Susceptibility to Degradation: A Global Assessment of the Problem and Options for Management, United Nations Environment Programme, Nairobi, Kenya, 2003.

5.   USCB, Statistical Abstract of the United States 2009, United States Census Bureau, Washington, DC, USA, 2009.

6.   DOE, Linking Legacies, DOE/EM-319, United States Department of Energy, Washington, DC, USA, 1997.

7.   K. S. Kasprzak, "Oxidative DNA and protein damage in metal-induced toxicity and carcinogenesis,"Free Radical Biology and Medicine, vol. 32, no. 10, pp. 958–967, 2002.

8.   S. Fukuda, "Chelating agents used for plutonium and uranium removal in radiation emergency medicine," Current Medicinal Chemistry, vol. 12, no. 23, pp. 2765–2770, 2005.

9.   M. Valko, H. Morris, and M. T. D. Cronin, "Metals, toxicity and oxidative stress," Current Medicinal Chemistry, vol. 12, no. 10, pp. 1161–1208, 2005.

10.   A. Galanis, A. Karapetsas, and R. Sandaltzopoulos, "Metal-induced carcinogenesis, oxidative stress and hypoxia signalling," Mutation Research—Genetic Toxicology and Environmental Mutagenesis, vol. 674, no. 1-2, pp. 31–35, 2009.

11.   J. C. Philp, S. M. Bamforth, I. Singleton, and R. M. Atlas, "Environmental pollution and restoration: a role for bioremediation," in Bioremediation: Applied Microbial Solutions for Real-World Environmental Cleanup, R. M. Atlas and J. Philp, Eds., pp. 1–48, ASM Press, Washington, DC, USA, 2005.

12.   EPA, "Search Superfund Site Information," 2013,http://cumulis.epa.gov/supercpad/cursites/srchsites.cfm.

13.   R. G. Riley, J. M. Zachara, and F. J. Wobber, Chemical Comtaminants on DOE Lands and Selection of Contamination Mixtures for Subsurface Science Research, DOE/ER-0547T, Energy UDo, Washington, DC, USA, 1992.

14.   C. Tamponnet, A. Martin-Garin, M.-A. Gonze et al., "An overview of BORIS: bioavailability of radionuclides in soils," Journal of Environmental Radioactivity, vol. 99, no. 5, pp. 820–830, 2008.

15.   C. Papastefanou, "Escaping radioactivity from coal-fired power plants (CPPs) due to coal burning and the associated hazards: a review," Journal of Environmental Radioactivity, vol. 101, no. 3, pp. 191–200, 2010.

16.   M. R. Palmer and J. M. Edmond, "Uranium in river water," Geochimica et Cosmochimica Acta, vol. 57, no. 20, pp. 4947–4955, 1993.

17. C. F. Jove Colon, P. V. Brady, M. D. Siegel, and E. R. Lindgren, "Historical case analysis of uranium plume attenuation," Soil and Sediment Contamination, vol. 10, no. 1, pp. 71–115, 2001.

18. T. M. Esat and Y. Yokoyama, "Variability in the uranium isotopic composition of the oceans over glacial-interglacial timescales," Geochimica et Cosmochimica Acta, vol. 70, no. 16, pp. 4140–4150, 2006.

19. S. Regenspurg, D. Schild, T. Schäfer, F. Huber, and M. E. Malmström, "Removal of uranium(VI) from the aqueous phase by iron(II) minerals in presence of bicarbonate," Applied Geochemistry, vol. 24, no. 9, pp. 1617–1625, 2009.

20. D. Langmuir, Aqueous Environmental Geochemistry, Prentice Hall, Upper Saddle River, NJ, USA, 1997.

21. A. S. Knox, M. H. Paller, D. D. Reible, X. Ma, and I. G. Petrisor, "Sequestering agents for active caps—remediation of metals and organics," Soil and Sediment Contamination, vol. 17, no. 5, pp. 516–532, 2008.

22. C. N. Mulligan, R. N. Yong, and B. F. Gibbs, "An evaluation of technologies for the heavy metal remediation of dredged sediments," Journal of Hazardous Materials, vol. 85, no. 1-2, pp. 145–163, 2001.

23. S. Chellam and D. A. Clifford, "Physical-chemical treatment of groundwater contaminated by leachate from surface disposal of uranium tailings," Journal of Environmental Engineering, vol. 128, no. 10, pp. 942–952, 2002.

24. W.-M. Wu, J. Carley, M. Fienen et al., "Pilot-scale in situ bioremediation of uranium in a highly contaminated aquifer. 1. Conditioning of a treatment zone," Environmental Science & Technology, vol. 40, no. 12, pp. 3978–3985, 2006.

25. T. Cheng, M. O. Barnett, E. E. Roden, and J. Zhuang, "Effects of phosphate on uranium(VI) adsorption to goethite-coated sand," Environmental Science and Technology, vol. 38, no. 22, pp. 6059–6065, 2004.

26. Y. Z. Tang and R. J. Reeder, "Uranyl and arsenate cosorption on aluminum oxide surface," Geochimica et Cosmochimica Acta, vol. 73, no. 10, pp. 2727–2743, 2009.

27. M. Vidal, M. J. Santos, T. Abrão, J. Rodríguez, and A. Rigol, "Modeling competitive metal sorption in a mineral soil," Geoderma, vol. 149, no. 3-4, pp. 189–198, 2009.

28. F. Z. El Aamrani, L. Duro, J. de Pablo, and J. Bruno, "Experimental study and modeling of the sorption of uranium(VI) onto olivine-rock," Applied Geochemistry, vol. 17, no. 4, pp. 399–408, 2002.

29. W. Luo, S. D. Kelly, K. M. Kemner et al., "Sequestering uranium and technetium through co-precipitation with aluminum in a contaminated acidic environment," Environmental Science and Technology, vol. 43, no. 19, pp. 7516–7522, 2009.

30. G. J. Vazquez, C. J. Dodge, and A. J. Francis, "Interaction of uranium(VI) with phthalic acid," Inorganic Chemistry, vol. 47, no. 22, pp. 10739–10743, 2008.

31. G. J. Vazquez, C. J. Dodge, and A. J. Francis, "Interactions of uranium with polyphosphate,"Chemosphere, vol. 70, no. 2, pp. 263–269, 2007.

32. T. Ozaki, T. Kimura, T. Ohnuki et al., "Association of europium(III), americium(III), and curium(III) with cellulose, chitin, and chitosan," Environmental Toxicology and Chemistry, vol. 25, no. 8, pp. 2051–2058, 2006.

33. C. Noubactep, A. Schöner, and G. Meinrath, "Mechanism of uranium removal from the aqueous solution by elemental iron," Journal of Hazardous Materials, vol. 132, no. 2-3, pp. 202–212, 2006.

34. F. G. Simon, V. Biermann, and B. Peplinski, "Uranium removal from groundwater using hydroxyapatite," Applied Geochemistry, vol. 23, no. 8, pp. 2137–2145, 2008.

35. R. D. Ludwig, D. J. A. Smyth, D. W. Blowes et al., "Treatment of arsenic, heavy metals, and acidity using a mixed ZVI-compost PRB," Environmental Science and Technology, vol. 43, no. 6, pp. 1970–1976, 2009.

36. Q. Y. Chen, M. Tyrer, C. D. Hills, X. M. Yang, and P. Carey, "Immobilisation of heavy metal in cement-based solidification/stabilisation: a review," Waste Management, vol. 29, no. 1, pp. 390–403, 2009.

37. S. Paria and P. K. Yuet, "Solidification-stabilization of organic and inorganic contaminants using portland cement: a literature

review," Environmental Reviews, vol. 14, no. 4, pp. 217–255, 2006.

38. A. S. Barinov, G. A. Varlakova, S. V. Stefanovskii, and M. I. Ozhovan, "Change of structure and properties of vitrified radioactive wastes during long-time storage in an experimental repository,"Atomic Energy, vol. 105, no. 2, pp. 110–117, 2008.

39. P. A. Bingham and R. J. Hand, "Vitrification of toxic wastes: a brief review," Advances in Applied Ceramics, vol. 105, no. 1, pp. 21–31, 2006.

40. H. Boukhalfa, S. D. Reilly, and M. P. Neu, "Complexation of Pu(IV) with the natural siderophore desferrioxamine B and the redox properties of Pu(IV)(siderophore) complexes," Inorganic Chemistry, vol. 46, no. 3, pp. 1018–1026, 2007.

41. R. J. Reeder, M. A. A. Schoonen, and A. Lanzirotti, "Metal speciation and its role in bioaccessibility and bioavailability," Reviews in Mineralogy and Geochemistry, vol. 64, pp. 59–113, 2006.

42. C. E. Halim, S. A. Short, J. A. Scott, R. Amal, and G. Low, "Modelling the leaching of Pb, Cd, As, and Cr from cementitious waste using PHREEQC," Journal of Hazardous Materials, vol. 125, no. 1–3, pp. 45–61, 2005.

43. J. R. Haas, T. J. Dichristina, and R. Wade Jr., "Thermodynamics of U(VI) sorption onto Shewanella putrefaciens," Chemical Geology, vol. 180, no. 1–4, pp. 33–54, 2001.

44. G. W. Bryan and W. J. Langston, "Bioavailability, accumulation and effects of heavy metals in sediments with special reference to United Kingdom estuaries," Environmental Pollution, vol. 76, no. 2, pp. 89–131, 1992.

45. W.-M. Wu, J. Carley, J. Luo et al., "In situ bioreduction of uranium (VI) to submicromolar levels and reoxidation by dissolved oxygen," Environmental Science and Technology, vol. 41, no. 16, pp. 5716–5723, 2007.

46. B. Faybishenko, T. C. Hazen, P. E. Long et al., "In situ long-term reductive bioimmobilization of Cr(VI) in groundwater using hydrogen release compound," Environmental Science & Technology, vol. 42, no. 22, pp. 8478–8485, 2008.

47.   S. Hong-Bo, C. Li-Ye, R. Cheng-Jiang, L. Hua, G. Dong-Gang, and L. Wei-Xiang, "Understanding molecular mechanisms for improving phytoremediation of heavy metal-contaminated soils," Critical Reviews in Biotechnology, vol. 30, no. 1, pp. 23–30, 2010.

48.   B. Van Aken, P. A. Correa, and J. L. Schnoor, "Phytoremediation of polychlorinated biphenyls: new trends and promises," Environmental Science and Technology, vol. 44, no. 8, pp. 2767–2776, 2010.

49.   D. R. Lovley, E. J. P. Phillips, Y. A. Gorby, and E. R. Landa, "Microbial reduction of uranium," Nature, vol. 350, no. 6317, pp. 413–416, 1991.

50.   L. E. Macaskie, R. M. Empson, A. K. Cheetham, C. P. Grey, and A. J. Skarnulis, "Uranium bioaccumulation by a Citrobacter sp. as a result of enzymically mediated growth of polycrystalline HUO_2PO_4," Science, vol. 257, no. 5071, pp. 782–784, 1992.

51.   A. J. Francis, C. J. Dodge, F. Lu, G. P. Halada, and C. R. Clayton, "XPS and XANES studies of uranium reduction by Clostridium sp.," Environmental Science Technology, vol. 28, no. 4, pp. 636–639, 1994.

52.   H. Pan-Hou, M. Kiyono, H. Omura, T. Omura, and G. Endo, "Polyphosphate produced in recombinantEscherichia coli confers mercury resistance," FEMS Microbiology Letters, vol. 207, no. 2, pp. 159–164, 2002.

53.   T. Barkay, S. M. Miller, and A. O. Summers, "Bacterial mercury resistance from atoms to ecosystems,"FEMS Microbiology Reviews, vol. 27, no. 2-3, pp. 355–384, 2003.

54.   A. Malik, "Metal bioremediation through growing cells," Environment International, vol. 30, no. 2, pp. 261–278, 2004.

55.   Y. Suzuki and J. F. Banfield, "Resistance to, and accumulation of, uranium by bacteria from a uranium-contaminated site," Geomicrobiology Journal, vol. 21, no. 2, pp. 113–121, 2004.

56.   T. Tsuruta, "Cell-associated adsorption of thorium or uranium from aqueous system using various microorganisms," Water, Air, and Soil Pollution, vol. 159, no. 1, pp. 35–47, 2004.

57.   R. J. Martinez, M. J. Beazley, M. Taillefert, A. K. Arakaki, J. Skolnick, and P. A. Sobecky, "Aerobic uranium (VI) bioprecipitation by

metal-resistant bacteria isolated from radionuclide- and metal-contaminated subsurface soils," Environmental Microbiology, vol. 9, no. 12, pp. 3122–3133, 2007.

58. R. A. Sanford, Q. Wu, Y. Sung et al., "Hexavalent uranium supports growth of Anaeromyxobacter dehalogenans and Geobacter spp. with lower than predicted biomass yields," Environmental Microbiology, vol. 9, no. 11, pp. 2885–2893, 2007.

59. M. L. Merroun and S. Selenska-Pobell, "Bacterial interactions with uranium: an environmental perspective," Journal of Contaminant Hydrology, vol. 102, no. 3-4, pp. 285–295, 2008.

60. M. J. Beazley, R. J. Martinez, P. A. Sobecky, S. M. Webb, and M. Taillefert, "Nonreductive biomineralization of uranium(VI) phosphate via microbial phosphatase activity in anaerobic conditions," Geomicrobiology Journal, vol. 26, no. 7, pp. 431–441, 2009.

61. G. M. Gadd, "Biosorption: critical review of scientific rationale, environmental importance and significance for pollution treatment," Journal of Chemical Technology and Biotechnology, vol. 84, no. 1, pp. 13–28, 2009.

62. J. O. Nriagu, "Phosphate minerals," in Phosphate Minerals: Their Properties and General Modes of Occurrence, J. O. Nriagu and P. B. Moore, Eds., pp. 1–137, Springer, Berlin, Germany, 1984.

63. V. Smil, "Phosphorus in the environment: natural flows and human interferences," Annual Review of Energy and the Environment, vol. 25, pp. 53–88, 2000.

64. K. B. Föllmi, "The phosphorus cycle, phosphogenesis and marine phosphate-rich deposits," Earth-Science Reviews, vol. 40, no. 1-2, pp. 55–124, 1996.

65. J. Rakovan, "Growth and surface properties of apatite," Reviews in Mineralogy and Geochemistry, vol. 48, no. 1, pp. 51–86, 2002.

66. S. L. McGowen, N. T. Basta, and G. O. Brown, "Use of diammonium phosphate to reduce heavy metal solubility and transport in smelter-contaminated soil," Journal of Environmental Quality, vol. 30, no. 2, pp. 493–500, 2001.

67. X. D. Cao, L. Q. Ma, M. Chen, S. P. Singh, and W. G. Harris, "Impacts of phosphate amendments on lead biogeochemistry at a contaminated site," Environmental Science and Technology, vol. 36, no. 24, pp. 5296–5304, 2002.

68. T. T. Eighmy and J. D. Eusden Jr., "Phosphate stabilization of municipal solid waste combustion residues: geochemical principles," Geological Society Special Publications, vol. 236, pp. 435–473, 2004.

69. V. S. Mehta, F. Maillot, Z. Wang, J. G. Catalano, and D. E. Giammar, "Effect of co-solutes on the products and solubility of uranium(VI) precipitated with phosphate," Chemical Geology, vol. 364, pp. 66–75, 2014.

70. J. S. Arey, J. C. Seaman, and P. M. Bertsch, "Immobilization of uranium in contaminated sediments by hydroxyapatite addition," Environmental Science & Technology, vol. 33, no. 2, pp. 337–342, 1999.

71. J. C. Seaman, J. S. Arey, and P. M. Bertsch, "Immobilization of nickel and other metals in contaminated sediments by hydroxyapatite addition," Journal of Environmental Quality, vol. 30, no. 2, pp. 460–469, 2001.

72. Y. Pan and M. E. Fleet, "Compositions of the apatite-group minerals: Substitution mechanisms and controlling factors," Reviews in Mineralogy and Geochemistry, vol. 48, no. 1, pp. 13–49, 2002.

73. A. S. Knox, D. I. Kaplan, D. C. Adriano, T. G. Hinton, and M. D. Wilson, "Apatite and phillipsite as sequestering agents for metals and radionuclides," Journal of Environmental Quality, vol. 32, no. 2, pp. 515–525, 2003.

74. E. S. Shelobolina, H. Konishi, H. Xu, and E. E. Roden, "U(VI) sequestration in hydroxyapatite produced by microbial glycerol 3-phosphate metabolism," Applied and Environmental Microbiology, vol. 75, no. 18, pp. 5773–5778, 2009.

75. D. R. Brookshaw, R. A. D. Pattrick, J. R. Lloyd, and D. J. Vaughan, "Microbial effects on mineral-radionuclide interactions and radionuclide solid-phase capture processes," Mineralogical Magazine, vol. 76, no. 3, pp. 777–806, 2012.

76. I. Llorens, G. Untereiner, D. Jaillard, B. Gouget, V. Chapon, and M. Carriere, "Uranium interaction with two multi-resistant environmental bacteria: Cupriavidus metallidurans CH34 and Rhodopseudomonas palustris," PLoS ONE, vol. 7, no. 12, Article ID e51783, 2012.

77. S. P. Singh, L. Q. Ma, and W. G. Harris, "Heavy metal interactions with phosphatic clay: sorption and desorption behavior," Journal of Environmental Quality, vol. 30, no. 6, pp. 1961–1968, 2001.

78. J. Yang, D. E. Mosby, S. W. Casteel, and R. W. Blanchar, "Lead immobilization using phosphoric acid in a smelter-contaminated urban soil," Environmental Science and Technology, vol. 35, no. 17, pp. 3553–3559, 2001.

79. W. D. Bostick, Use of Apatite for Chemical Stabilization of Subsurface Contaminants, U.S. Department of Energy, Washington, DC, USA, 2003.

80. F. G. Simon, V. Biermann, C. Segebade, and M. Hedrich, "Behaviour of uranium in hydroxyapatite-bearing permeable reactive barriers: Investigation using ^{237}U as a radioindicator," Science of the Total Environment, vol. 326, no. 1–3, pp. 249–256, 2004.

81. Y. J. Lee, E. J. Elzinga, and R. J. Reeder, "Sorption mechanisms of zinc on hydroxyapatite: systematic uptake studies and EXAFS spectroscopy analysis," Environmental Science and Technology, vol. 39, no. 11, pp. 4042–4048, 2005.

82. D. M. Wellman, J. P. Icenhower, and A. T. Owen, "Comparative analysis of soluble phosphate amendments for the remediation of heavy metal contaminants: effect on sediment hydraulic conductivity," Environmental Chemistry, vol. 3, no. 3, pp. 219–224, 2006.

83. R. Q. Liu and D. Y. Zhao, "In situ immobilization of Cu(II) in soils using a new class of iron phosphate nanoparticles," Chemosphere, vol. 68, no. 10, pp. 1867–1876, 2007.

84. D. M. Wellman, E. M. Pierce, and M. M. Valenta, "Efficacy of soluble sodium tripolyphosphate amendments for the in-situ immobilisation of uranium," Environmental Chemistry, vol. 4, no. 5, pp. 293–300, 2007.

85. A. Hwang, W. Ji, B. Kweon, and J. Khim, "The physico-chemical properties and leaching behaviors of phosphatic clay for immobilizing heavy metals," Chemosphere, vol. 70, no. 6, pp. 1141–1145, 2008.

86. W. R. Berti and S. D. Cunningham, "In-place inactivation of Pb in Pb-contaminated soils," Environmental Science and Technology, vol. 31, no. 5, pp. 1359–1364, 1997.

87.   S. Brown, R. Chaney, J. Hallfrisch, J. A. Ryan, and W. R. Berti, "In situ soil treatments to reduce the phyto- and bioavailability of lead, zinc, and cadmium," Journal of Environmental Quality, vol. 33, no. 2, pp. 522–531, 2004.

88.   M. J. A. Rijkenberg and C. V. Depree, "Heavy metal stabilization in contaminated road-derived sediments," Science of the Total Environment, vol. 408, no. 5, pp. 1212–1220, 2010.

89.   K. G. Scheckel and J. A. Ryan, "Spectroscopic speciation and quantification of lead in phosphate-amended soils," Journal of Environmental Quality, vol. 33, no. 4, pp. 1288–1295, 2004.

90.   M. Chrysochoou, D. Dermatas, and D. G. Grubb, "Phosphate application to firing range soils for Pb immobilization: the unclear role of phosphate," Journal of Hazardous Materials, vol. 144, no. 1-2, pp. 1–14, 2007.

91.   M. F. Fanizza, H. Yoon, C. Zhang et al., "Pore-scale evaluation of uranyl phosphate precipitation in a model groundwater system," Water Resources Research, vol. 49, no. 2, pp. 874–890, 2013.

92.   K. L. Nash, M. P. Jensen, and M. A. Schmidt, "Actinide immobilization in the subsurface environment by in-situ treatment with a hydrolytically unstable organophosphorus complexant: uranyl uptake by calcium phytate," Journal of Alloys and Compounds, vol. 271–273, pp. 257–261, 1998.

93.   J. C. Seaman, J. M. Hutchison, B. P. Jackson, and V. M. Vulava, "In situ treatment of metals in contaminated soils with phytate," Journal of Environmental Quality, vol. 32, no. 1, pp. 153–161, 2003.

94.   C. De Stefano, D. Milea, N. Porcino, and S. Sammartano, "Speciation of phytate ion in aqueous solution. Sequestering ability toward mercury(II) cation in $NaCl_{aq}$ at different ionic strengths," Journal of Agricultural and Food Chemistry, vol. 54, no. 4, pp. 1459–1466, 2006.

95.   C. De Stefano, G. Lando, D. Milea, A. Pettignano, and S. Sammartano, "Formation and stability of cadmium(II)/Phytate complexes by different electrochemical techniques. Critical analysis of results,"Journal of Solution Chemistry, vol. 39, no. 2, pp. 179–195, 2010.

96. F. Crea, C. De Stefano, D. Milea, and S. Sammartano, "Formation and stability of phytate complexes in solution," Coordination Chemistry Reviews, vol. 252, no. 10-11, pp. 1108–1120, 2008.

97. Q. Y. Ma, T. J. Logan, and S. J. Traina, "Lead immobilization from aqueous solutions and contaminated soils using phosphate rocks," Environmental Science and Technology, vol. 29, no. 4, pp. 1118–1126, 1995.

98. W. D. Bostick, R. J. Stevenson, R. J. Jarabek, and J. L. Conca, "Use of apatite and bone char for the removal of soluble radionuclides in authentic and simulated DOE groundwater," Advances in Environmental Research, vol. 3, pp. 488–498, 1999.

99. M. E. Hodson, E. Valsami-Jones, and J. D. Cotter-Howells, "Bonemeal additions as a remediation treatment for metal contaminated soil," Environmental Science and Technology, vol. 34, no. 16, pp. 3501–3507, 2000.

100. C. C. Fuller, J. R. Bargar, and J. A. Davis, "Molecular-scale characterization of uranium sorption by bone apatite materials for a permeable reactive barrier demonstration," Environmental Science and Technology, vol. 37, no. 20, pp. 4642–4649, 2003.

101. S. B. Chen, Y. G. Zhu, and Y. B. Ma, "The effect of grain size of rock phosphate amendment on metal immobilization in contaminated soils," Journal of Hazardous Materials, vol. 134, no. 1–3, pp. 74–79, 2006.

102. J. K. Yoon, X. Cao, and L. Q. Ma, "Application methods affect phosphorus-induced lead immobilization from a contaminated soil," Journal of Environmental Quality, vol. 36, no. 2, pp. 373–378, 2007.

103. P. Thakur, R. C. Moore, and G. R. Choppin, "Sorption of U(VI) species on hydroxyapatite,"Radiochimica Acta, vol. 93, no. 7, pp. 385–391, 2005.

104. S. Raicevic, J. V. Wright, V. Veljkovic, and J. L. Conca, "Theoretical stability assessment of uranyl phosphates and apatites: selection of amendments for in situ remediation of uranium," Science of the Total Environment, vol. 355, no. 1-3, pp. 13–24, 2006.

105. J. Oliva, J. de Pablo, J.-L. Cortina, J. Cama, and C. Ayora, "Removal of cadmium, copper, nickel, cobalt and mercury from water by Apatite II: column experiments," Journal of Hazardous Materials, vol. 194, pp. 312–323, 2011.

106. W. D. Schecher and D. C. McAvoy, "MINEQL+: a software environment for chemical equilibrium modeling," Computers, Environment and Urban Systems, vol. 16, no. 1, pp. 65–76, 1992.

107. R. Guillaumont, T. Fanghänel, J. Fuger, et al., "Chemical thermodynamics 5," in Update on the Chemical Thermodynamics of Uranium, Neptunium, Plutonium, Americium and Technetium, F. J. Mompean, M. Illemassene, C. Domenech-Orti, and K. Ben-Said, Eds., Elsevier, Amsterdam, The Netherlands, 2003.

108. R. Q. Liu and D. Y. Zhao, "Reducing leachability and bioaccessibility of lead in soils using a new class of stabilized iron phosphate nanoparticles," Water Research, vol. 41, no. 12, pp. 2491–2502, 2007.

109. Z. Z. Zhang, M. Y. Li, W. Chen, S. Z. Zhu, N. N. Liu, and L. Zhu, "Immobilization of lead and cadmium from aqueous solution and contaminated sediment using nano-hydroxyapatite," Environmental Pollution, vol. 158, no. 2, pp. 514–519, 2010.

110. E. Navarro, A. Baun, R. Behra et al., "Environmental behavior and ecotoxicity of engineered nanoparticles to algae, plants, and fungi," Ecotoxicology, vol. 17, no. 5, pp. 372–386, 2008.

111. M. Motskin, D. M. Wright, K. Muller et al., "Hydroxyapatite nano and microparticles: correlation of particle properties with cytotoxicity and biostability," Biomaterials, vol. 30, no. 19, pp. 3307–3317, 2009.

112. X. Zhao, S. Ng, B. C. Heng et al., "Cytotoxicity of hydroxyapatite nanoparticles is shape and cell dependent," Archives of Toxicology, vol. 87, no. 6, pp. 1037–1052, 2013.

113. D. Voet and J. G. Voet, Biochemistry, John Wiley & Sons, Hoboken, NJ, USA, 2004.

114. C. P. Vance, C. Uhde-Stone, and D. L. Allan, "Phosphorus acquisition and use: critical adaptations by plants for securing a nonrenewable resource," New Phytologist, vol. 157, no. 3, pp. 423–447, 2003.

115. K. Y. Kim, D. Jordan, and H. B. Krishnan, "Rahnella aquatilis, a bacterium isolated from soybean rhizosphere, can solubilize hydroxyapatite," FEMS Microbiology Letters, vol. 153, no. 2, pp. 273–277, 1997.

116. G. M. Rossolini, S. Schippa, M. L. Riccio, F. Berlutti, L. E. Macaskie, and M. C. Thaller, "Bacterial nonspecific acid phosphohydrolases: physiology, evolution and use as tools in microbial biotechnology,"Cellular and Molecular Life Sciences, vol. 54, no. 8, pp. 833–850, 1998.

117. K. Y. Kim, D. Jordan, and H. B. Krishnan, "Expression of genes from Rahnella aquatilis that are necessary for mineral phosphate solubilization in Escherichia coli," FEMS Microbiology Letters, vol. 159, no. 1, pp. 121–127, 1998.

118. M. A. Whitelaw, T. J. Harden, and K. R. Helyar, "Phosphate solubilisation in solution culture by the soil fungus penicillium radicum," Soil Biology and Biochemistry, vol. 31, no. 5, pp. 655–665, 1999.

119. O. A. Vershinina and L. V. Znamenskaya, "The Pho regulons of bacteria," Microbiology, vol. 71, no. 5, pp. 497–511, 2002.

120. F. D. Dakora and D. A. Phillips, "Root exudates as mediators of mineral acquisition in low-nutrient environments," Plant and Soil, vol. 245, no. 1, pp. 35–47, 2002.

121. H. Lambers, M. W. Shane, M. D. Cramer, S. J. Pearse, and E. J. Veneklaas, "Root structure and functioning for efficient acquisition of phosphorus: matching morphological and physiological traits,"Annals of Botany, vol. 98, no. 4, pp. 693–713, 2006.

122. G. M. Gadd, "Geomycology: biogeochemical transformations of rocks, minerals, metals and radionuclides by fungi, bioweathering and bioremediation," Mycological Research, vol. 111, no. 1, pp. 3–49, 2007.

123. M. Ben Farhat, A. Farhat, W. Bejar et al., "Characterization of the mineral phosphate solubilizing activity of Serratia marcescens CTM 50650 isolated from the phosphate mine of Gafsa," Archives of Microbiology, vol. 191, no. 11, pp. 815–824, 2009.

124. S. Uroz, C. Calvaruso, M. P. Turpault et al., "Efficient mineral weathering is a distinctive functional trait of the bacterial genus Collimonas," Soil Biology & Biochemistry, vol. 41, no. 10, pp. 2178–2186, 2009.

125. A. Kornberg, N. N. Rao, and D. Ault-Riché, "Inorganic polyphosphate: a molecule of many functions,"Annual Review of Biochemistry, vol. 68, pp. 89–125, 1999.

126. J. W. McGrath, S. Cleary, A. Mullan, and J. P. Quinn, "Acid-stimulated phosphate uptake by activated sludge microorganisms under aerobic laboratory conditions," Water Research, vol. 35, no. 18, pp. 4317–4322, 2001.

127. M. J. Seufferheld, H. M. Alvarez, and M. E. Farias, "Role of polyphosphates in microbial adaptation to extreme environments," Applied and Environmental Microbiology, vol. 74, no. 19, pp. 5867–5874, 2008.

128. T. Shiba, K. Tsutsumi, H. Yano et al., "Inorganic polyphosphate and the induction of rpoS expression," Proceedings of the National Academy of Sciences of the United States of America, vol. 94, no. 21, pp. 11210–11215, 1997.

129. H. Gonzalez and T. E. Jensen, "Nickel sequestering by polyphosphate bodies in Staphylococcus aureus," Microbios, vol. 93, no. 376, pp. 179–185, 1998.

130. K. Tsutsumi, M. Munekata, and T. Shiba, "Involvement of inorganic polyphosphate in expression of SOS genes," Biochimica et Biophysica Acta: Gene Structure and Expression, vol. 1493, no. 1-2, pp. 73–81, 2000.

131. P. L. Foster, "Stress-induced mutagenesis in bacteria," Critical Reviews in Biochemistry and Molecular Biology, vol. 42, pp. 373–397, 2007.

132. L. Andrade, C. N. Keim, M. Farina, and W. C. Pfeiffer, "Zinc detoxification by a cyanobacterium from a metal contaminated bay in Brazil," Brazilian Archives of Biology and Technology, vol. 47, no. 1, pp. 147–152, 2004.

133. N. Renninger, R. Knopp, H. Nitsche, D. S. Clark, and J. D. Keasling, "Uranyl precipitation by Pseudomonas aeruginosa via controlled polyphosphate metabolism," Applied and Environmental Microbiology, vol. 70, no. 12, pp. 7404–7412, 2004.

134. M. Kiyono and H. Pan-Hou, "Genetic engineering of bacteria for environmental remediation of mercury," Journal of Health Science, vol. 52, no. 3, pp. 199–204, 2006.

135. F. Remonsellez, A. Orell, and C. A. Jerez, "Copper tolerance of the thermoacidophilic archaeon Sulfolobus metallicus: possible role of polyphosphate metabolism," Microbiology, vol. 152, no. 1, pp. 59–66, 2006.

136. N. Perdrial, N. Liewig, J.-E. Delphin, and F. Elsass, "TEM evidence for intracellular accumulation of lead by bacteria in subsurface environments," Chemical Geology, vol. 253, no. 3-4, pp. 196–204, 2008.

137. T. J. Reilly, G. S. Baron, F. E. Nano, and M. S. Kuhlenschmidt, "Characterization and sequencing of a respiratory burst-inhibiting acid phosphatase from Francisella tularensis," The Journal of Biological Chemistry, vol. 271, no. 18, pp. 10973–10983, 1996.

138. B. L. Wanner, "Phosphorus assimilation and control of the phosphate regulon," in Escherichia coli and Salmonella, F. C. Neidhardt, R. Curtiss III, and J. L. Ingraham, Eds., vol. 1 of Cellular and Molecular Biology, pp. 1357–1381, ASM Press, Washington, DC, USA, 2nd edition, 1996.

139. A. M. Burroughs, K. N. Allen, D. Dunaway-Mariano, and L. Aravind, "Evolutionary genomics of the HAD superfamily: understanding the structural Adaptations and catalytic diversity in a superfamily of phosphoesterases and allied enzymes," Journal of Molecular Biology, vol. 361, no. 5, pp. 1003–1034, 2006.

140. J. B. Vincent, M. W. Crowder, and B. A. Averill, "Hydrolysis of phosphate monoesters: a biological problem with multiple chemical solutions," Trends in Biochemical Sciences, vol. 17, no. 3, pp. 105–110, 1992.

141. H. G. Hoppe, "Phosphatase activity in the sea," Hydrobiologia, vol. 493, pp. 187–200, 2003.

142. V. N. Anupama, P. N. Amrutha, G. S. Chitra, and B. Krishnakumar, "Phosphatase activity in anaerobic bioreactors for wastewater treatment," Water Research, vol. 42, no. 10-11, pp. 2796–2802, 2008.

143. G. M. Gadd, "Metals, minerals and microbes: geomicrobiology and bioremediation," Microbiology, vol. 156, no. 3, pp. 609–643, 2010.

144. L. E. Macaskie and A. C. R. Dean, "Cadmium accumulation by micro-organisms," Environmental Technology Letters, vol. 3, no. 2, pp. 49–56, 1982.

145. L. E. Macaskie and A. C. R. Dean, "Strontium accumulation by immobilized cells of a Citrobacter sp.,"Biotechnology Letters, vol. 7, no. 9, pp. 627–630, 1985.

146. B. C. Jeong, P. S. Poole, A. C. Willis, and L. E. Macaskie, "Purification and chacterization of acid-type phosphatases from a heavy- metal-accumulating Citrobacter sp.," Archives of Microbiology, vol. 169, no. 2, pp. 166–173, 1998.

147. M. Paterson-Beedle, L. E. Macaskie, C. H. Lee, J. A. Hriljac, K. Y. Jee, and W. H. Kim, "Utilisation of a hydrogen uranyl phosphate-based ion exchanger supported on a biofilm for the removal of cobalt, strontium and caesium from aqueous solutions," Hydrometallurgy, vol. 83, no. 1–4, pp. 141–145, 2006.

148. R. L. Felts, T. J. Reilly, M. J. Calcutt, and J. J. Tanner, "Cloning, purification and crystallization of Bacillus anthracis class C acid phosphatase," Acta Crystallographica Section F, vol. 62, no. 7, pp. 705–708, 2006.

149. L. G. Powers, H. J. Mills, A. V. Palumbo, C. Zhang, K. Delaney, and P. A. Sobecky, "Introduction of a plasmid-encoded phoA gene for constitutive overproduction of alkaline phosphatase in three subsurfacePseudomonas isolates," FEMS Microbiology Ecology, vol. 41, no. 2, pp. 115–123, 2002.

150. D. Appukuttan, A. S. Rao, and S. K. Apte, "Engineering of Deinococcus radiodurans R1 for bioprecipitation of uranium from dilute nuclear waste," Applied and Environmental Microbiology, vol. 72, no. 12, pp. 7873–7878, 2006.

151. K. S. Nilgiriwala, A. Alahari, A. S. Rao, and S. K. Apte, "Cloning and overexpression of alkaline phosphatase PhoK from Sphingomonas sp. strain BSAR-1 for bioprecipitation of uranium from alkaline solutions," Applied and Environmental Microbiology, vol. 74, no. 17, pp. 5516–5523, 2008.

152. A. Geissler, M. Merroun, G. Geipel, H. Reuther, and S. Selenska-Pobell, "Biogeochemical changes induced in uranium mining waste pile samples by uranyl nitrate treatments under anaerobic conditions," Geobiology, vol. 7, no. 3, pp. 282–294, 2009.

153. M. J. Beazley, R. J. Martinez, S. M. Webb, P. A. Sobecky, and M. Taillefert, "The effect of pH and natural microbial phosphatase activity on the speciation of uranium in subsurface soils," Geochimica et Cosmochimica Acta, vol. 75, no. 19, pp. 5648–5663, 2011.

154. A. L. Neal, K. Lowe, T. L. Daulton, J. Jones-Meehan, and B. J. Little, "Oxidation state of chromium associated with cell surfaces

of Shewanella oneidensis during chromate reduction," Applied Surface Science, vol. 202, no. 3-4, pp. 150–159, 2002.

155. P. Pattanapipitpaisal, A. N. Mabbett, J. A. Finlay et al., "Reduction of Cr(VI) and bioaccumulation of chromium by gram positive and gram negative microorganisms not previously exposed to Cr-stress,"Environmental Technology, vol. 23, no. 7, pp. 731–745, 2002.

156. C. E. Mire, J. A. Tourjee, W. F. O'Brien, K. V. Ramanujachary, and G. B. Hecht, "Lead precipitation byVibrio harveyi : evidence for novel quorum-sensing interactions," Applied and Environmental Microbiology, vol. 70, no. 2, pp. 855–864, 2004.

157. M. J. Beazley, R. J. Martinez, P. A. Sobecky, S. M. Webb, and M. Taillefert, "Uranium biomineralization as a result of bacterial phosphatase activity: insights from bacterial isolates from a contaminated subsurface," Environmental Science & Technology, vol. 41, no. 16, pp. 5701–5707, 2007.

158. F. Jroundi, M. L. Merroun, J. M. Arias, A. Rossberg, S. Selenska-Pobell, and M. T. González-Muñoz, "Spectroscopic and microscopic characterization of uranium biomineralization in Myxococcus xanthus,"Geomicrobiology Journal, vol. 24, no. 5, pp. 441–449, 2007.

159. J. de Pablo, I. Casas, J. Giménez et al., "The oxidative dissolution mechanism of uranium dioxide. I. The effect of temperature in hydrogen carbonate medium," Geochimica et Cosmochimica Acta, vol. 63, no. 19-20, pp. 3097–3103, 1999.

160. K. T. Finneran, M. E. Housewright, and D. R. Lovley, "Multiple influences of nitrate on uranium solubility during bioremediation of uranium-contaminated subsurface sediments," Environmental Microbiology, vol. 4, no. 9, pp. 510–516, 2002.

161. J. K. Fredrickson, J. M. Zachara, D. W. Kennedy et al., "Influence of Mn oxides on the reduction of uranium(VI) by the metal-reducing bacterium Shewanella putrefaciens," Geochimica et Cosmochimica Acta, vol. 66, no. 18, pp. 3247–3262, 2002.

162. C. X. Liu, J. M. Zachara, J. K. Fredrickson, D. W. Kennedy, and A. Dohnalkova, "Modeling the inhibition of the bacterial reduction of U(VI) by β-MnO$_2$(s)," Environmental Science and Technology, vol. 36, no. 7, pp. 1452–1459, 2002.

163. J. D. Istok, J. M. Senko, L. R. Krumholz et al., "In situ bioreduction of technetium and uranium in a nitrate-contaminated aquifer.," Environmental Science and Technology, vol. 38, no. 2, pp. 468–475, 2004.

164. B. Gu, H. Yan, P. Zhou, D. B. Watson, M. Park, and J. Istok, "Natural humics impact uranium bioreduction and oxidation," Environmental Science and Technology, vol. 39, no. 14, pp. 5268–5275, 2005.

165. J. Wan, T. K. Tokunaga, E. Brodie et al., "Reoxidation of bioreduced uranium under reducing conditions," Environmental Science and Technology, vol. 39, no. 16, pp. 6162–6169, 2005.

166. M. Ginder-Vogel, C. S. Criddle, and S. Fendorf, "Thermodynamic constraints on the oxidation of biogenic UO_2 by Fe(III) (Hydr) oxides," Environmental Science and Technology, vol. 40, no. 11, pp. 3544–3550, 2006.

167. J. D. Wall and L. R. Krumholz, "Uranium reduction," Annual Review of Microbiology, vol. 60, pp. 149–166, 2006.

168. H. S. Moon, J. Komlos, and P. R. Jaffé, "Uranium reoxidation in previously bioreduced sediment by dissolved oxygen and nitrate," Environmental Science and Technology, vol. 41, no. 13, pp. 4587–4592, 2007.

169. B. Gu, W.-M. Wu, M. A. Ginder-Vogel et al., "Bioreduction of uranium in a contaminated soil column," Environmental Science and Technology, vol. 39, no. 13, pp. 4841–4847, 2005.

170. W.-M. Wu, J. Carley, T. Gentry et al., "Pilot-scale in situ bioremedation of uranium in a highly contaminated aquifer. 2. Reduction of U(VI) and geochemical control of U(VI) bioavailability," Environmental Science and Technology, vol. 40, no. 12, pp. 3986–3995, 2006.

171. R. S. Oremland and J. F. Stolz, "The ecology of arsenic," Science, vol. 300, no. 5621, pp. 939–944, 2003.

172. R. P. Deo and B. E. Rittmann, "A biogeochemical framework for bioremediation of plutonium(V) in the subsurface environment," Biodegradation, vol. 23, no. 4, pp. 525–534, 2012.

173. R. Finch and T. Murakami, "Systematics and paragenesis of uranium minerals," in Uranium: Mineralogy, Geochemistry and the Environment, P. C. Burns and R. Finch, Eds., Mineralogical Society of America, Washington, Wash, USA, 1999.

174. E. C. Buck, N. R. Brown, and N. L. Dietz, "Contaminant uranium phases and leaching at the Fernald site in Ohio," Environmental Science & Technology, vol. 30, no. 1, pp. 81–88, 1996.

175. Y. Roh, S. R. Lee, S.-K. Choi, M. P. Elless, and S. Y. Lee, "Physicochemical and mineralogical characterization of uranium-contaminated soils," Soil and Sediment Contamination, vol. 9, no. 5, pp. 463–486, 2000.

176. C. C. Fuller, J. R. Bargar, J. A. Davis, and M. J. Piana, "Mechanisms of uranium interactions with hydroxyapatite: implications for groundwater remediation," Environmental Science & Technology, vol. 36, no. 2, pp. 158–165, 2002.

177. D. M. Wellman, J. N. Glovack, K. Parker, E. L. Richards, and E. M. Pierce, "Sequestration and retention of uranium(VI) in the presence of hydroxylapatite under dynamic geochemical conditions,"Environmental Chemistry, vol. 5, no. 1, pp. 40–50, 2008.

178. J. R. Lloyd, P. Yong, and L. E. Macaskie, "Biological reduction and removal of Np(V) by two microorganisms," Environmental Science & Technology, vol. 34, no. 7, pp. 1297–1301, 2000.

179. M. I. Boyanov, K. E. Fletcher, M. J. Kwon et al., "Solution and microbial controls on the formation of reduced U(IV) species," Environmental Science & Technology, vol. 45, no. 19, pp. 8336–8344, 2011.

180. D. K. Newman and J. F. Banfield, "Geomicrobiology: how molecular-scale interactions underpin biogeochemical systems," Science, vol. 296, no. 5570, pp. 1071–1077, 2002.

181. H. L. Ehrlich, "Geomicrobiology: its significance for geology," Earth Science Reviews, vol. 45, no. 1-2, pp. 45–60, 1998.

182. J. Macalady and J. F. Banfield, "Molecular geomicrobiology: genes and geochemical cycling," Earth and Planetary Science Letters, vol. 209, no. 1-2, pp. 1–17, 2003.

183. EPA, "Use of monitored natural attenuation at superfund, RCRA corrective action and underground storage tank sites," OSWER Directive 9200.4-17P, U.S. Environmental Protection Agency, Washington, Wash, USA, 1999.

184. J. Fruchter, "In situ treatment of chromium-contaminated groundwater," Environmental Science & Technology, vol. 36, no. 23, pp. 464A–472A, 2002.

185. C. N. Mulligan and R. N. Yong, "Natural attenuation of contaminated soils," Environment International, vol. 30, no. 4, pp. 587–601, 2004.

186. EPA, Monitored Natural Attenuation of Inorganic Contaminants in Ground Water Volume 1—Technical Basis for Assessment, EPA/600/R-07/139, U.S. Environmental Protection Agency, Washington, DC, USA, 2007.

187. EPA, "Monitored natural attenuation of inorganic contaminants in ground water, Volume 2—Assessment for non-radionuclides including arsenic, cadmium, chromium, copper, lead, nickel, nitrate, perchlorate, and selenium," EPA/600/R-07/140, Environmental Protection Agency, Washington, DC, USA, 2007.

188. G. Basnakova and L. E. Macaskie, "Microbially-enhanced chemisorption of Ni^{2+} ions into biologically-synthesised hydrogen uranyl phosphate (HUP) and selective recovery of concentrated Ni^{2+} using citrate or chloride ion," Biotechnology Letters, vol. 23, no. 1, pp. 67–70, 2001.

189. J. E. Stubbs, D. C. Elbert, D. R. Veblen, and C. Zhu, "Electron microbeam investigation of uranium-contaminated soils from Oak Ridge, TN, USA," Environmental Science and Technology, vol. 40, no. 7, pp. 2108–2113, 2006.

190. K. R. Salome, S. J. Green, M. J. Beazley, S. M. Webb, J. E. Kostka, and M. Taillefert, "The role of anaerobic respiration in the immobilization of uranium through biomineralization of phosphate minerals," Geochimica et Cosmochimica Acta, vol. 106, pp. 344–363, 2013.

191. M. Fomina, J. M. Charnock, S. Hillier, R. Alvarez, and G. M. Gadd, "Fungal transformations of uranium oxides," Environmental Microbiology, vol. 9, no. 7, pp. 1696–1710, 2007.

192. T. Reitz, M. L. Merroun, A. Rossberg, and S. Selenska-Pobell, "Interactions of Sulfolobus acidocaldariuswith uranium," Radiochimica Acta, vol. 98, no. 5, pp. 249–257, 2010.

193. T. Reitz, M. L. Merroun, A. Rossberg, R. Steudtner, and S. Selenska-Pobell, "Bioaccumulation of U(VI) by Sulfolobus acidocaldarius under moderate acidic conditions," Radiochimica Acta, vol. 99, no. 9, pp. 543–553, 2011.

194. E. L. Brodie, T. Z. DeSantis, D. C. Joyner et al., "Application of a high-density oligonucleotide microarray approach to study bacterial population dynamics during uranium reduction and reoxidation," Applied and Environmental Microbiology, vol. 72, no. 9, pp. 6288–6298, 2006.

195. J. D. Van Nostrand, W.-M. Wu, L. Wu et al., "GeoChip-based analysis of functional microbial communities during the reoxidation of a bioreduced uranium-contaminated aquifer," Environmental Microbiology, vol. 11, no. 10, pp. 2611–2626, 2009.

196. I. Porat, T. A. Vishnivetskaya, J. J. Mosher et al., "Characterization of archaeal community in contaminated and uncontaminated surface stream sediments," Microbial Ecology, vol. 60, no. 4, pp. 784–795, 2010.

197. G. Rastogi, S. Osman, P. A. Vaishampayan, G. L. Andersen, L. D. Stetler, and R. K. Sani, "Microbial diversity in uranium mining-impacted soils as revealed by high-density 16S microarray and clone library," Microbial Ecology, vol. 59, no. 1, pp. 94–108, 2010.

198. T. M. Gihring, G. Zhang, C. C. Brandt et al., "A limited microbial consortium is responsible for extended bioreduction of uranium in a contaminated aquifer," Applied and Environmental Microbiology, vol. 77, no. 17, pp. 5955–5965, 2011.

199. J. D. Van Nostrand, L. Wu, W.-M. Wu et al., "Dynamics of microbial community composition and function during in situ bioremediation of a uranium-contaminated aquifer," Applied and Environmental Microbiology, vol. 77, no. 11, pp. 3860–3869, 2011.

200. K. Katsaveli, D. Vayenas, G. Tsiamis, and K. Bourtzis, "Bacterial diversity in Cr(VI) and Cr(III)-contaminated industrial wastewaters," Extremophiles, vol. 16, no. 2, pp. 285–296, 2012.

201. Y. Liang, J. D. Van Nostrand, L. A. N'Guessan et al., "Microbial functional gene diversity with a shift of subsurface redox conditions during in situ uranium reduction," Applied and Environmental Microbiology, vol. 78, no. 8, pp. 2966–2972, 2012.

202. K. Chourey, S. Nissen, and T. Vishnivetskaya, "Environmental proteomics reveals early microbial community responses to biostimulation at a uranium- and nitrate-contaminated site," Proteomics, vol. 13, no. 18-19, pp. 2921–2930, 2013.

203. K. M. Handley, N. C. VerBerkmoes, C. I. Steefel et al., "Biostimulation induces syntrophic interactions that impact C, S and N cycling in a sediment microbial community," ISME Journal, vol. 7, no. 4, pp. 800–816, 2013.

204. S. Kang, J. D. Van Nostrand, H. L. Gough et al., "Functional gene array-based analysis of microbial communities in heavy metals-contaminated lake sediments," FEMS Microbiology Ecology, vol. 86, pp. 200–214, 2013.

205. A. C. Somenahally, J. J. Mosher, T. Yuan, et al., "Hexavalent chromium reduction under fermentative conditions with lactate stimulated native microbial communities," Plos ONE, vol. 8, no. 12, Article ID e83909, 2013.

206. R. J. Martinez, C. H. Wu, and M. J. Beazley, "Microbial community responses to organophosphate substrate additions in contaminated subsurface sediments," Plos ONE, vol. 9, no. 6, Article ID e100383, 2014.

207. S. D. Kelly, K. M. Kemner, J. B. Fein et al., "X-ray absorption fine structure determination of pH-dependent U-bacterial cell wall interactions," Geochimica et Cosmochimica Acta, vol. 66, no. 22, pp. 3855–3871, 2002.

208. S. D. Kelly, K. M. Kemner, J. Carley et al., "Speciation of uranium in sediments before and after in situbiostimulation," Environmental Science & Technology, vol. 42, no. 5, pp. 1558–1564, 2008.

209. S. D. Kelly, W.-M. Wu, F. Yang et al., "Uranium transformations in static microcosms," Environmental Science and Technology, vol. 44, no. 1, pp. 236–242, 2010.

210. DOE, "New frontiers in characterizing biological systems: report from the May 2009 workshop," Tech. Rep. DOE/SC-0121, U.S. Department of Energy Office of Science, Washington, DC, USA, 2009.

Chapter 8

Heavy Metal Distributions in Water of the Aras River, Ardabil, Iran

Fatemeh Nasehi[1], Amirhesam Hassani[1], Masoud Monavvari[1], Abdoreza Karbassi[1], Nematollah Khorasani[1], and Aliakbar Imani[2]

[1]Department of Environment and Energy, Science and Research Branch, Islamic Azad University, Tehran, Iran

[2]Department Agronomy and Plant Breeding, Islamic Azad University, Ardabil Branch, Ardabil, Iran

ABSTRACT

Aras (situated on the frontier river) is one of Iran's important rivers which is situated on the west North of the country. The concentration of heavy metals in this river was studied around Ardabil province during the 1389 in all four seasons (spring, summer, autumn, winter). The cluster analysis technique with the help of the results gained from density of metals like (Zn, Cu, Fe, Hg, Ni, Pb, Cd) was used in the water of Aras river for classifying quality of the river. According to the gained results from the cluster analysis, the stations were divided into three groups with high pollution (HP), medium pollution (MP) and low pollution (LP). In general, S3, S5 stations with high pollution, S2, S4 stations with medium pollution and S1 station with low pollutions are classified in the water of Aras River.

INTRODUCTION

About 60 km of Aras River is situated in north part of Ardabil province, between Aslandouz to Tazeh kand in Parsabad.

Creating deviation band of Mill Moghan and following that establishing irrigation networks in Moghan has caused this region to be one of the main agricultural poles of the country.

So every kind of agricultural activity is done in this area. In general, nowadays, the most important pollution sources, overlooking Aras river is in Ardabil, including: agriculture, city and village waste waters, rubbish, Industrial waste water (industrial town of parsabad) and marine nourishment that have been situated at different parts of this wide area. In recent years, with everyday development of human activities in the countries around Aras river, the probability of ecosystem changes is not an odd idea [1,2]. Interring chemical damages into area are often because of a mixture of natural and anthropogenic matters [3]. Industrial and agricultural waste water, city waste water, mines and extra materials of heavy metals are evacuated into rivers and seas at last. A part of metal as solution, apart as suspended sediment and a part as river—bed, is carried and entered the area [4]. The metals in the first two phases, are available and before being as sediment are moved easily and enter the food web. If high dose of a heavy metal enter the food web, a live creatures can react to the availability of it in different ways [5-7]. Because of this point, concentration of some of heavy metals in Greenland, sea creatures has been studied and the results showed that the density of these elements in sea creatures is more than the density of land ecosystem and more important that they found density of (Hg, Fe, Cd) in higher food levels increases, it means biological accumulation happens [8]. Also, According to these results, the relation between diseases wide spreading and density of heavy metals has been studied [1,9].

Since, Aras River is one of the rare and various habitat of the country. Any kind of pollution in it will affect the existent variety of the river. On the other hand, this river water is used for agriculturing and norishing the fish of the province [10]. Thus, for estimating the amount of heavy metals entrance to the river and finally to the water of seas and lakes, measuring and estimating heavy metals concentration in the water of river is essential. In this point, for studying the condition of river pollution and spotting the affect of polluting sources an it, the amount of heavy metals in the Aras water in different stations around Ardabil province, (from the entrance to the province to its exit) was studied. In this research, seven heavy metals (Hg, Cd, Zn, Cu, Fe, Pb) have been studied in the water of Aras river.

THE STUDY REGION

Aras River is a frontier river with the length of 1072 km, which 460 km of it is at the common borderline of Iran and Armenia and Azerbaijan [11]. Aras is as the vital river of the region and the main source for Moghan irrigation. From the slope of Bina Gul Dagh Mountains, Turkey source, and the total water shed is 100,220 square km.

In normal years, the maximum of the measured debi in Aras is about 1100 cub meter per second in dam which is used as a tank (north of Makou) and 2600 cub meter per second in deviant dam of Mill and Moghan, that these numbers in dry years, change respectively to 32 and 180 cub meter per second.

THE METHOD

For choosing a situation for sampling, besides providing topography and geology maps, the most important sources of pollutions and the exit ways of their waste water that are ending to Aras River, has been studied. Regarding the above mentioned cases and thinking of the ways to the river (For its being a frontier river) five stations were chosen for sampling. The station 1, Mill Moghan dam (the entrance of Aras into Ardabil), station 2 after Darre roud joining to Aras, Station 3 after Gouri chay joining to Aras, station 4 (Oltan) Parsabad industrial town shoal and station 5 (Tazeh kand village) the exit of Aras from Iran (Figure 1).

Water sample were chosen from the depth of 20 cm under water and were put in polyetilen dishes [12]. Immediately after collecting water sample, the suspended materials were separated by using special filters which one good for separating articles bigger than 45 percent micrometer, because the suspended and colloidal articles affect the chemical analysis results of water [2].

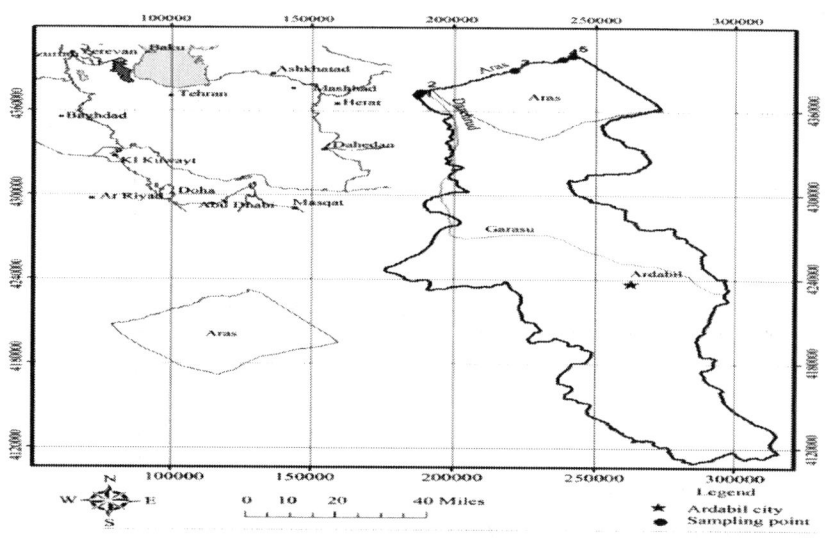

Figure 1: Sampling stations within the study area.

After straining the water in the operation place, water got acidic. Measuring metals in the water of four seasons, spring, summer, autumn and winter, was done. The sample of spring was done in the first half of Jun, summer in the second half of August, autumn in the second half of November and winter in the second half of January.

Analysis

Measuring zinc, iron, lead, copper and cadmium in water samples were in this order that first we add 3 cc Acid Nitric without mercury into the water of 200 cc inside a dish. For avoiding mistake after cooling the sample, Again we add 3 cc thick Acid Nitric with a mild heat and after digesting 2 cc added Acid Nitric (1 + 1), We rise it up to the volume of 25 cc. the metal concentration is read by the atomic absorption spectrophotometer [13]. Measuring Nickel and mercury was done too after digestion processes by ICP [14].

Statistic Processes

The analysis of given articles was done by using cluster analysis [6]. The main goal of cluster analysis is classify the stations according to their features [15]. In a way that stations with like features are put in one group [16]. In this research in order to cluster analysis, after standardizing the given data, are used these stages based on distance [17] and ward method based on the least square Euclidean distance [18]. Dendron gram cut based on the farthest regional distance, divided the stations into three groups.

Concentration of Heavy Metals in Water

Table one shows the density of metals copper, zinc, Iron, lead, Nickel, Mercury and cadmium in the water of Aras river in 5 chosen stations and in different four season, spring, summer, autumn and winter.

The mean of copper density in spring is 70 microgram per one liter with the limitation of the density about 7 - 101 microgram per liter, which their maximum has been measured in station 3 and 5. The mean of copper density in summer is 100 microgram per liter that its maximum is present in station 2. The mean of copper density in autumn and winter is respectively 11 and 8. The mean of zinc density in spring is 38 microgram per liter and its maximum is about 118 microgram per liter in stations 3 and it's mean in summer is 72 microgram and with the maximum of 144 microgram per liter is shown is station 2,which in this area Darre roud is lead into Aras. The copper density mean is 67 microgram per liter in autumn and its maximum is 135 microgram per liter in station 5 (Tazeh kand). The order of seasons according to density reduction is: summer, autumn, spring, winter.

The mean of Iron density is spring is 177 microgram per liter and its maximum is shown in station 2. The mean of Iron density in summer is 102 microgram per liter and its maximum is in station 2. Iron density reduction in seasons is: spring-summer-autumn-winter. The mean of Nickel density in spring is 44 microgram per liter and its maximum is in station 3 with the amount of 71 microgram per liter.

The mean of Nickel in summer is 38 microgram per liter and its maximum is in station 5 (Tazeh kand village), with the amount of 80 microgram per liter. Nickel density reduction is: spring-summer-autumn and winter. The mean of lead density in spring is 8 microgram per liter and its maximum is in station 2 (after joining of Darre roud to Aras) with the amount of 12 microgram per liter.

The mean of lead density in summer is 5 microgram per liter and has the same amount in all stations. The order of lead density reduction in water is: spring-summer-autumn-winter. The mean of cadmium density in spring is 8 microgram per liter and in summer with the highest amount of 9 microgram per liter and its maximum is in station 2.

The amount of the density of same metals such as Nickel and lead and Iron has a direct relation with the amount of rain and river water, and with a high probability, in the seasons with high amount of water. They enter the river because of soil erosion and river bed dissolution. Second, in spring the farmer start their fight against the fungicides. Mixture of copper and other metals are in fungicides and they can be one of the reasons of their increase in the river in spring. The reason of the high density of some metals in summer rather than spring can be because of river water reduction in this season and following that, can be the reduction of water surface and the suspended materials in it.

When the load of the river reduces, the metal absorption reduces from solution phase into dissolution phase too. In winter, Because of the reduction of poison and fertilizer and reduction of other industrial activities, the pollution load is reduced too. Besides, in winter because of temperate decrease, metal dissolving in water decreases too [19]. Finally, the amount of metals is another source of copper and other metals in the environment, because some mixture such as copper sulphate for controlling blights. Of course, existence of copper in a high amount (is scarcely 0.05 microgram in normal) is for human activities. Pesticide poisons are used for confronting the insects and fertilizer in summer. And since mine fertilizers have high amount of copper (0.01 - 0.05 microgram per liter) the high amount of copper in these stations can be related to the use of fertilizer and poison in irrigation network in Moghan. Nickel is a heavy metal that exists in main industrial polluting areas for sediments. The high amount of Nickel can be found in cereal, beaned fruit, and soybean productions [20] that existence of factories for fruit juice, demise, oil cake, livestock food and sugar in this area, it means in territory of Pars abad overlooking Aras River can be one of the reasons for the increase of this element in water. The waste water of houses of the city which are entering the river, is one of the ways of distribution of Nickel in water [21].

RESULTS OF CLUSTER ANALYSIS

The results from Table 1 and dendrogram in water samples shows that the first cluster includes stations S3, S5, these two stations are in one cluster and have similar densities of metals. Station 3 is the accepter of the pollution of Gouri chay in to Aras. In this station, besides the pollution of stations 1 and 2, the pollutions of Guri chay also enter to Aras River, And also the waste water of Aslandouz directly enter this region. The station 5, the shoal of industrial waste water 15 industrial unit is working. Moghan cultivation waste water (near oltan and old Tazeh kand), sugar, dairies and domesticated food of parsabad which are one of the most important and biggest industrial units in this area and one of dangerous biological centers are also entering this station.

The second cluster includes stations S2, S4 that the pollution of station 2 is after joinery of Darre rud river into Aras. Darre rud river bring the industrial, agricultural waste water of village and towns of Tabriz, Oroumieh and Armenia into Aras. Mines are one of the entrance ways of metals such as copper into the water since there are some copper mines from Azerbaijan and Armenia.

Station 5 like station 4 also gets the industrial waste water of Parsabad. The third cluster includes station 1 which shows low densities of metals rather than the others.

Table 1: The concentration of heavy metals in Aras River during 4 seasons (the units are microgram per liter ppb)

Station Name	season	Cu	Fe	Zn	Ni	Cd	Pb	Hg
	spring	7	65	5	58	ND	8	ND
S1	summer	7	29	32	24	ND	5	ND

	autumn	5	13	13	5	ND	7	ND
	winter	NI)	15	ND	ND	ND	ND	ND
	spring	80	299	26	41	7	12	ND
S2	summer	155	27	144	15	10	5	ND
	autumn	16	143	120	6	ND	7	ND
	winter	ND	39	5	6	ND	ND	ND
	spring	101	240	118	71	9	5	ND
S3	summer	112	90	35	32	9	5	ND
	autumn	7	35	6	6	ND	6	ND
	winter	10	5	ND	ND	ND	ND	ND
	spring	60	190	10	10	8	9	ND
S4	summer	95	103	46	41	8	5	ND
	autumn	8	47	62	10	5	8	ND
	winter	7	7	5	3	ND	5	ND
S5	spring	101	91	28	44	6	5	ND
	summer	104	64	105	80	7	5	ND
	autumn	19	64	135	28	8	9	ND
	winter	8	32	6	4	ND	5	ND

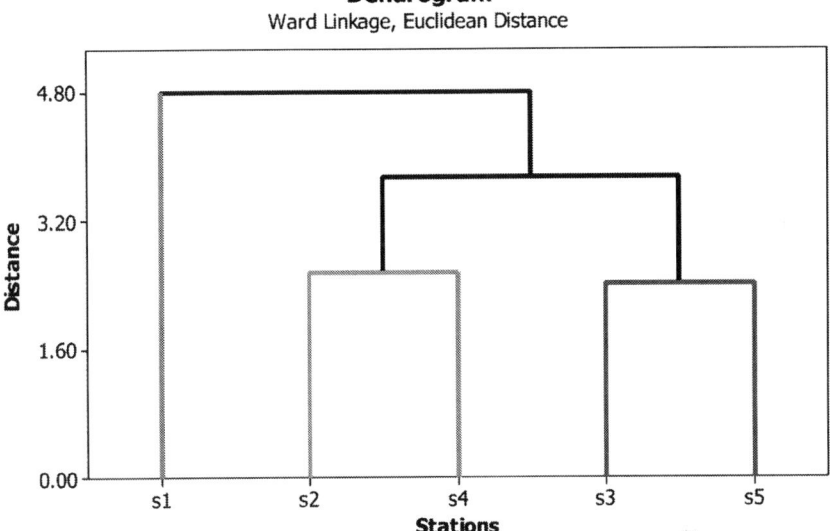

Figure 2: The results Dendrogram of the analysis of stations for heavy metals in Aras River.

CONCLUSIONS

According to the results of chart and Dendrogram in water samples, in general we can conclude that station 3 and 5 the highly polluted stations, among the others (HP), which are different by their pollution and exist farther than the others (Figure 2). After them, stations 2 and 4 with medium pollutions. Station1 is the lowest according to its pollution. In this research, it's shown that the differences between groups (stations) show the differences between polluting sources [15].

REFERENCES

1.   A. Baghvand, T. Nasrabadi, G. R. Nabi Bidhendi, A. Vosoogh, A. R. Karbassi and N. Mehrdadi, "Groundwater Quality Degradation of an Aquifer in Iran Central Desert," Desalination, Vol. 260, No. 1-3, 2010, pp. 264-275. doi:10.1016/j.desal.2010.02.038

2.   T. Nasrabadi, G. R. Nabi Bidhendi, A. R. Karbassi, H. Hoveidi, I. Nasrabadi, H. Pezeshk and F. Rashidinejad, "Influence of Sungun Copper Mine on Groundwater Quality, NW Iran," Environmental Geology, Vol. 58, No. 4, 2009, pp. 693-700. doi:10.1007/s00254-008-1543-2

3.   G. R. Nabi Bidhendi, A. R. Karbassi, T. Nasrabadi and H. Hoveidi, "Influence of Copper Mine on Surface Water Quality," International Journal of Environmental Science and Technology, Vol. 4, No. 1, 2007, pp. 85-91.

4.   R. Pardo, E. Barrado, L. Perez and M. Vega, "Determination and Association of Heavy Metals in Sediments of the Pisucrga, River," Water Research, Vol. 24, No. 3, 1990, pp. 373-379. doi:10.1016/0043-1354(90)90016-Y

5.   G. T. Ankley, D. M. Di Toro, D. J. Hansen and W. J. Berry, "Technical Basis and Proposal for Deriving Sediment Quality Criteria for Metals," Environmental Toxicology and Chemistry, Vol. 15, No. 12, 1996, pp. 2056- 2066. doi:10.1002/etc.5620151202

6.   K. P. Singh, A. Malik, D. Mohan and S. Sinha, "Water Quality Assessment and Apportionment of Pollution Sources of Gomti River (India) Using Multivariate Statistical Techniques: A Case Study," Analytica Chimica Acta, Vol. 538, No. 1-2, 2005, pp. 355-374. doi:10.1016/j.aca.2005.02.006

7.   V. K. Sharma, K. B. Rhhudy, R. Koening and F. G. Vazquez, "Metals in Sediments of the Upper Languna Madra," Marine Pollution Bulletin, Vol. 38, No. 12, 1999, pp. 1221-1226. doi:10.1016/S0025-326X(99)00166-6

8.   A. R. Karbassi, J. Nouri and G. O. Ayaz, "Flocculation of Trace Metals during Mixing of Talar River Water with Caspian Seawater," International Journal of Environmental Research, Vol. 1, No. 1, 2007, pp. 66-73.

9.   R. Chester and R. M. Hughes, "A Chemical Technique for the Separation of Ferro-Manganese Minerals, Carbonate Minerals and Adsorbed Trace Elements from Pelagic Sediment," Chemical Geology, Vol. 2, 1967, pp. 249-262. doi:10.1016/0009-2541(67)90025-3

10.   M. V. Farabi, et al., "Physical, Chemical, Biological Studies of Aras River Heave Metals," Fishery Research Institute of Iran, Khazar Sea Research House, 1388.

11. Z. A. Bagirov and S. E. Bravarnik, "Water Management and Power Use of the Araks River," Power Technology and Engineering, Vol. 19, No. 1, 2005, pp. 42-47.

12. USEPA, "Volunteer Stream Monitoring, a Method Manual for Water Quality Monitoring," 1997, p. 177.

13. M. Otto, "Multivariate Methods," In: R. Kellner, J. M. Mermet and H. M. Widmer, Eds., Analytical Chemistry, Wiley-VCH, Weinheim, 1998.

14. APHA, "Standard Methods for the Examination of Water and Wastewater," American Public Health Association, Washington, DC, 1992.

15. D. Carlton, S. W. Rust and L. Sinnott, "Application of Statistical Modeling to Optimize a Coastal Water Qulity Monitoring Program," Environmental Monitoring Assessment, Vol. 132, No. 1-3, 2007, pp. 505-522

16. M. Karamouz, F. Szidarovszky and B. Zahraie, "Water Resources Systems Analysis," Lewis, Washington, DC, 2003, pp. 369-386.

17. M. Matthies, J. Berlekamp, S. Lautenbach, N. Graf and S. Reimer, "System Analysis of Water Quality Management for the Elbe River Basin," Environmental Modeling & Software, Vol. 21, No. 5, 2006, pp. 1309-1318. doi:10.1016/j.envsoft.2005.04.026

18. S. Shrestha and F. Kazama, "Assessment of Surface Water Quality Using Multivariate Statistical Techniques: A Case Study of the Fuji River Basin," Japanese Environmental Modeling and Software, Vol. 22, No. 4, 2007, pp. 464-475.doi:10.1016/j. envsoft.2006.02.001

19. J. I. Drever, "The Geochemistry of Natural Waters," 3rd Edition, Prentice Hall Inc., Upper Saddle River, 1997, p. 436.

20. E. Merian, Ed., "Metals and Their Compounds in the Environment: Occurrence," Analysis and Biological Relevance, VCH, Weinheim, 1991, p. 1438.

21. M. Vega, R. Pardo, E. Barrado and L. Deban, "Assessment of Seasonal and Polluting Effects on the Quality of River Water by Exploratory Data Analysis," Water Research, Vol. 32, No. 12, 1998, pp. 3581-3592. doi:10.1016/S0043-1354(98)00138-9

9

Adaptive Neuro-Fuzzy Logic System for Heavy Metal Sorption in Aquatic Environments

Ahmad Qasaimeh[1], Mohammad Abdallah[2], and Falah Bani Hani[2]

[1]Department of Civil Engineering, Jerash University, Jerash, Jordan
[2]Chemical Engineering Department, AlHuson University College, Al-Balqa Applied University, Salt, Jordan

ABSTRACT

In this paper, adaptive neuro-fuzzy inference system ANFIS is used to assess conditions required for aquatic systems to serve as a sink for metal removal; it is used to generate information on the behavior of heavy metals (mercury) in water in relation to its uptake by bio-species (e.g. bacteria, fungi, algae, etc.) and adsorption to sediments. The approach of this research entails training fuzzy inference system by neural networks. The process is useful when there is interrelation between variables and no enough experience about mercury behavior, furthermore it is easy and fast process. Experimental work on mercury removal in wetlands for specific environmental conditions was previously conducted in bench scale at Concordia University laboratories. Fuzzy inference system FIS is constructed comprising knowledge base (i.e. premises and conclusions), fuzzy sets, and fuzzy rules. Knowledge base and rules are adapted and trained by neural networks, and then tested. ANFIS simulates and predicts mercury speciation for biological uptake and mercury adsorption to sediments. Modeling of mercury bioavailability for bio-species and adsorption to sediments shows strong correlation of more than 98% between simulation results and experimental data. The fuzzy models obtained are used to simulate and forecast further information on mercury partitioning to species and sediments. The findings of this research give information about metal removal by aquatic systems and their efficiency.

INTRODUCTION

The release of heavy metals from industries into the environment has resulted in many problems for both human health and aquatic ecosystems [1,2]. Heavy metals released into the environment by technological activities tend to persist indefinitely, circulating and eventually accumulating throughout the food chain, becoming a serious threat to the environment [3]. The presence of heavy metals in the environment is of major concern because of their toxicity, bio-accumulating tendency, threat to human life and the environment [4,5]. Lead, cadmium and mercury are examples of heavy metals that have been classified as priority pollutants by the U.S Environmental protection Agency (US EPA) [6]. Various biomaterials have been examined for their biosorptive properties and different types of biomass have shown levels of metal uptake [7]. Tables 1 and 2 show examples of biomass ability to sorb heavy metals.

In recent years, applying biotechnology in controlling and removing metal pollution has been paid much attention, and gradually becomes hot topic in the field of metal pollution control because of its potential application. Alternative process is biosorption, which utilizes various certain natural materials of biological origin, including bacteria, fungi, yeast, algae, plant, etc. These biosorbents possess metal-buffering property and can be used to decrease the concentration of heavy metal ions in solution [8]. A large quantity of materials has been investigated as biosorbents for the removal of metals extensively. The tested biosorbents can be basically classified into the following categories: bacteria (e.g. Bacillus subtillis), fungi (e.g. Rhizopus arrhizus) (Table 3), yeast (e.g., Saccharomyces cerevisiae), algae, industrial wastes (e.g., S. cerevisiae waste biomass from fermentation and food industry), water plants (e.g. Water Hyacinths and Reeds), agricultural wastes (e.g. corn core), and other polysaccharide materials [9].

The importance of metallic ions to fungal and yeast metabolism has been known for a long time [10]. The yeast biomass has been successfully used as biosorbent for removal of Ag, Au, Cd, Co, Cr, Cu, Ni, Pb, U, Th and Zn from aqueous solution. Yeasts of genera Saccharomyces, Candida, Pichia are efficient biosorbents for heavy metal ions [11].

Algae are of special interest in search for and the development of new biosorbents materials due to their high sorption capacity and their ready availability in practically unlimited quantities in the aquatic systems as seas and oceans (Table 4) [24,25].

Table 1: Biomass and their biosorbent capacity [12]

Type of biomass	Biosorbent capacity (meq/g)
Sargassum sp.	2 - 2.3
Rhizopus arrhizus	1.1
Peat moss	4.5 - 5.0
Eclonia radiate	1.8 - 2.4
Commercial resins	0.35 - 5.0

Table 2: Metal biosorption capacity by different biosorbents

Biosorbent	Metal	Capacity (mg/g)	Source
Bacillus sp.	Pb	92.2	[13]
Aeromonas caviae	Cd	155.3	[14]
Sargassum	Cu	56	[15]
Streptomyces rimosus	Fe(III)	122.0	[16]

Bacillus coagulans	Cr(IV)	39.9	[17]
Ulva reticulata	Cu	74.6	[18]
Bacillus thuringiensis	Ni	45.9	[19]
Ganoderma lucidum	Cu	24	[20]
Bacillus megaterium	Th	74.0	[21]
Sunflower stalk	Cu	29.3	[22]

Table 3: The value of sorption capacity (mmol/g) for different sorbents [23]

Sorbent	Cu	Pb	Cr
Pseudomonas aeruginosa (bacteria)	0.29	0.33	-
Rhizopus arrhizus (fungus)	0.20	0.50	0.27
Activated charcoal granular	0.03	0.15	0.07

Aquatic systems are considered as natural ecosystems that are designed to take advantage of the natural processes to provide efficient and low-cost wastewater treatment. The removal of metals from the water column within an aquatic system is performed generally by biological species uptake and by adsorption to sediments. Therefore, the aquatic systems such as streams, rivers, reservoirs and lakes serve as sink for heavy metal in aqueous solutions. The metal ion bioavailability for sorption to the biotic surface is pH dependent, as well for metal ion adsorption to sediments. The binding of a metal ion to the biotic surface of an organism decreases with increasing pH, whereas the binding behavior of metal ion to sediments increases with increasing pH.

METHODOLOGY

The research methodology implies Neuro-Fuzzy system to model and assess mercury removal from aquatic natural systems. Investigational data and information for mercury bioavailability in water and adsorption to sediments were adopted from previous research work-literature review of Prof. Elektorowicz research team attained in Concordia University-Canada [26-32].

Table 4: Algae sorption capacity (mmol/g) for metal ions in aqueous solution

Metal ion	Brown algae	Red algae	Green algae	Average Capacity
Cd	0.93	0.26	0.6	0.81
Ni	0.87	0.27	0.51	0.73
Zn	0.68	-	0.37	0.21
Cu	1.02	-	0.5	0.91
Pb	1.24	0.65	0.81	1.13

Neuro-Fuzzy system simulates mercury sorption and evaluates the efficiency of removal by verifying the effects of pH and mercury concentration in water. The computational tools used in this research are those in MATLAB; fuzzy toolbox and simulink.

ADAPTIVE NEURO-FUZZY INFERENCE SYSTEM (ANFIS)

In natural systems where variables are interrelated and data is large, it is difficult to determine the membership functions for input variables. Neuro-adaptive learning technique works similarly to that of neural networks. It provides a method for the fuzzy modeling procedure to learn information about a data set. Fuzzy Logic computes the membership function parameters that best allow the associated fuzzy inference system to track the given input/output data. The Fuzzy Logic accomplishes this membership function parameter adjustment is called adaptive neuro-fuzzy inference system. Using a given input/ output data set, the fuzzy inference system uses either a backpropagation algorithm alone or in combination with a least squares type of method. This adjustment allows fuzzy systems to learn from the data they are modeling. Then testing the data to check the generalization capability of the resulting fuzzy inference system is needed.

Checking the data is set for model validation. Model validation is the process by which the non-trained input variables are presented to the trained fuzzy inference system model to see how well the model predicts the corresponding output data.

Adaptive Neuro-Fuzzy Inference System for Mercury Speciation

The fuzzy inference system consists of two components: the linguistic term base (database) and the rule base. The database is fuzzified in two parts: fuzzy premises (input) and fuzzy conclusions (output). The fuzzy production rule base infers input to output and then defuzzified.

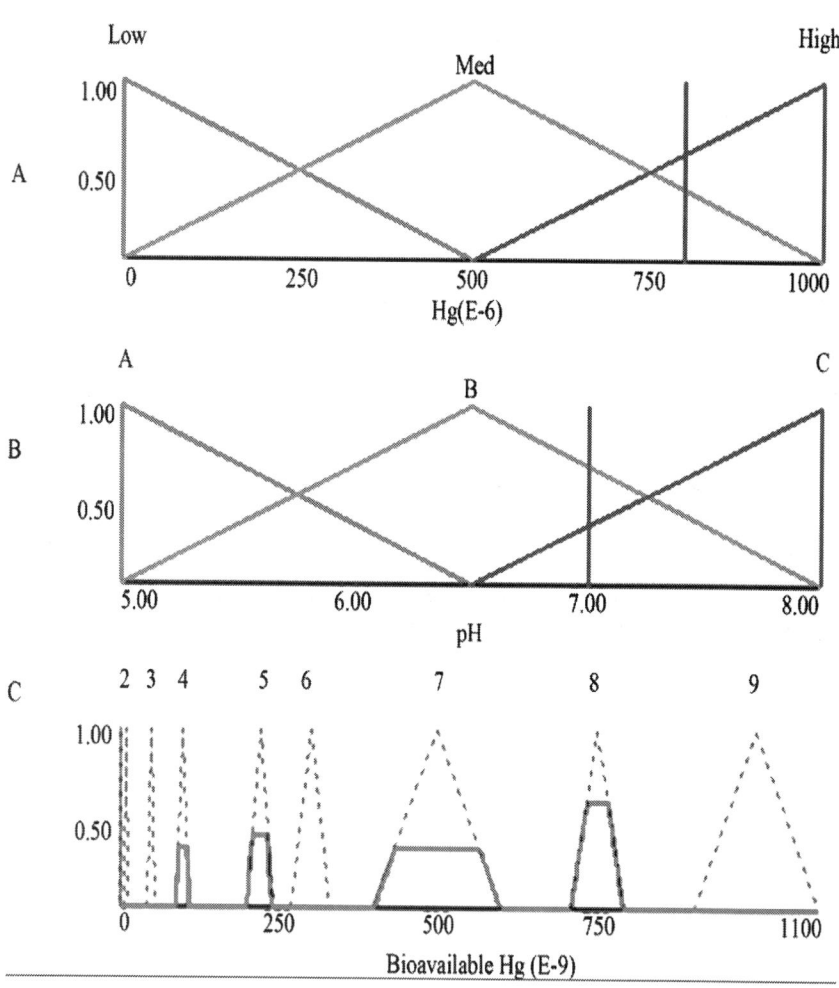

(a)

input   inputmf   rule   outputmf   output

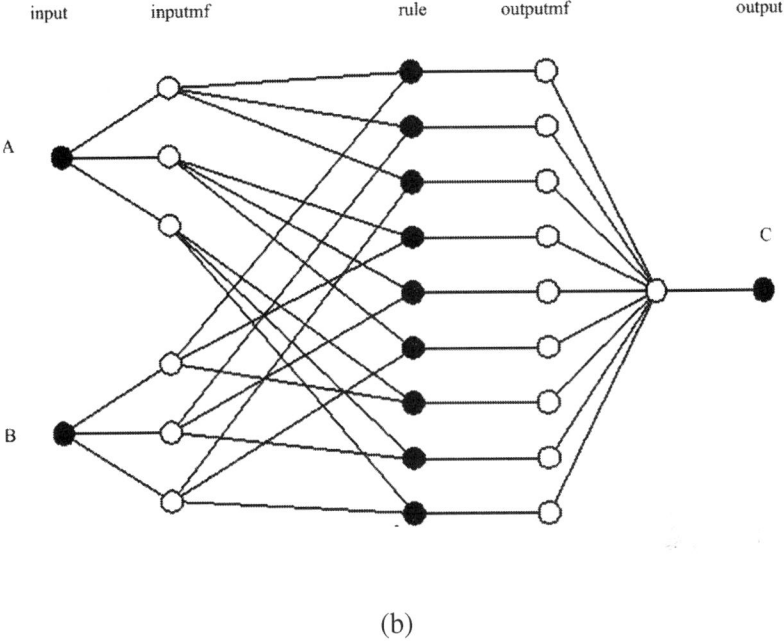

(b)

Figure 1: Fuzzy inference system.

Figure 1 shows a scheme for fuzzy inference system.

Adaptive Neuro-Fuzzy Inference System (ANFIS) is applied to estimate mercury removal in natural waters. ANFIS simulates and predicts mercury bioavailability that will be bio-sorbed by biological species. ANFIS model entails the following input variables to estimate output variable (bioavailable mercury concentration):

- Initial concentration of total mercury is in the range 0.3×10^{-6} - 1×10^{-3} moles/l;
- The pH value is situated in the range 5.36 - 8.

ANFIS model is constructed into two inputs (Hg_i and pH), one output (Bioavailable Hg), and nine rules. ANFIS model, training the data, and training error are illustrated in Figure 2.

ANFIS model fits the experimental data for bioavailable mercury concentration. Subsequently, comparison is conducted between results obtained from the model using ANFIS and experimental results for different initial total mercury concentrations in water and pH. The comparison shows strong correlation (Figures 3 and 4).

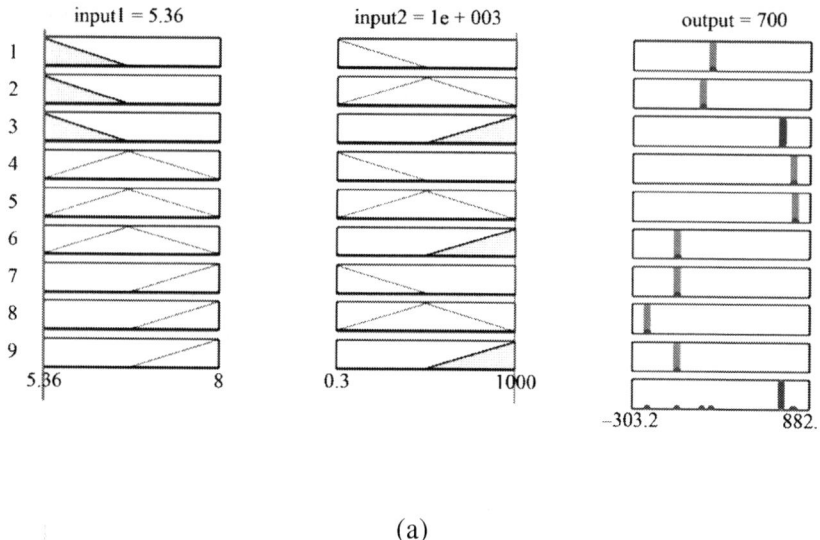

(a)

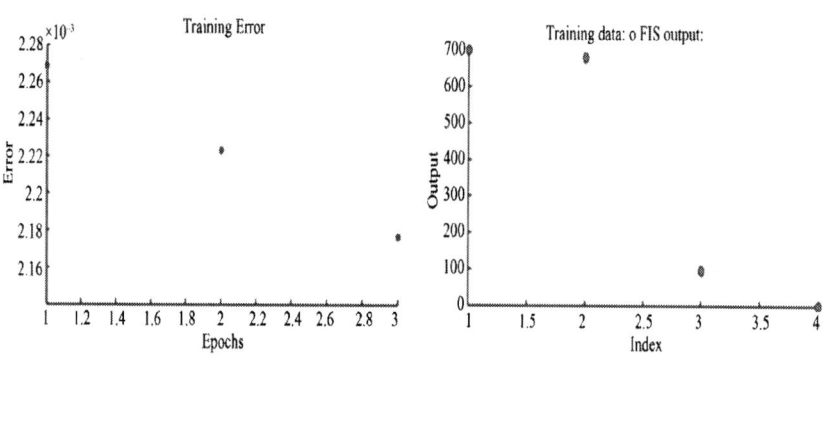

(b)

Figure 2: ANFIS model and training for mercury bioavailability in water by varying initial Hg concentration and pH.

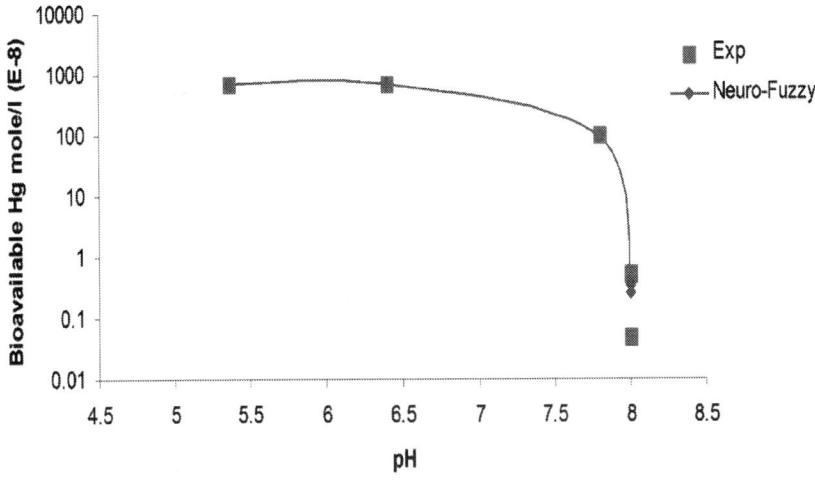

Figure 3: Simulation of bioavailable mercury concentration to be uptaken by bio-species verses pH for the range of initial mercury concentration.

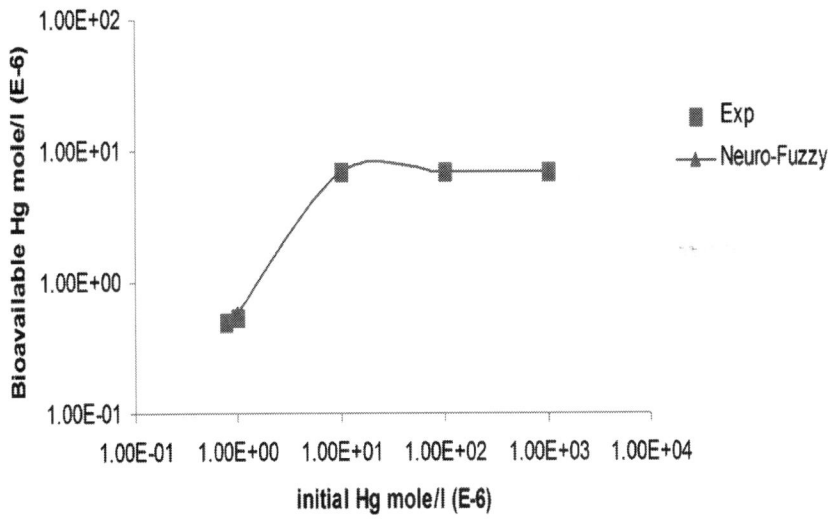

Figure 4: Simulation of bioavailable mercury concentration to be uptaken by bio-species versus initial total mercury concentrations and pH of 5.36.

Adaptive Neuro-Fuzzy System for Mercury Adsorption

In the second stage of work, ANFIS provides solution for soil adsorption of mercury for different conditions of initial mercury concentration and pH value. Neuro-Fuzzy system depends on fuzzy knowledge bases that satisfy the following parameters:

- The initial concentration of total mercury is in the range between 1×10^{-7} to 1×10^{-3} mole/l;
- The pH value is located between 5.36 to 8;
- The adsorbent concentration is 10 g/l.

ANFIS model is consisted of two inputs (Hg_i and pH), one output (adsorbed Hg), and nine rules. The model and training the data are shown in Figure 5. ANFIS model shows strong correlation between simulation and experimental data of mercury adsorption within different initial mercury concentrations and pH as shown in Figures 6 and 7.

For certain initial mercury concentrations, the adsorbed mercury is decreasing from its upper value when pH equals 8 to its lower value and when pH equals 5.36 as shown in Figure 6. The benefit of neural training of the fuzzy inference system is vital especially when there is large data and no experience of the system behavior.

SIMULATION AND FORECASTING

In previous sections the ANFIS model is constructed, trained, and checked. Now the model is ready for further range of simulation and forecasting. More information could be predicted for mercury removal by sorption in aquatic natural systems. In this section a simulink diagram is used for forecasting. Figure 8 shows an example of using fuzzy logic systems for mercury bioavailability and adsorption that were produced in previous sections to expand more information about Hg removal.

The run of simulink model describes the total mercury removal performance by components of an aquatic natural system; this performance can be depicted within different pH at certain initial mercury concentration. Figure 9 shows the performance of an aquatic system where it is optimal at pH equals 6.5. The fitting equation in the figure provides forecasting for total removal of mercury by natural waters components (bio-species and sediments) at any value of pH. The analysis in this section supply more information for model performance and forecasting. It also gives information about removal efficiency of the overall system.

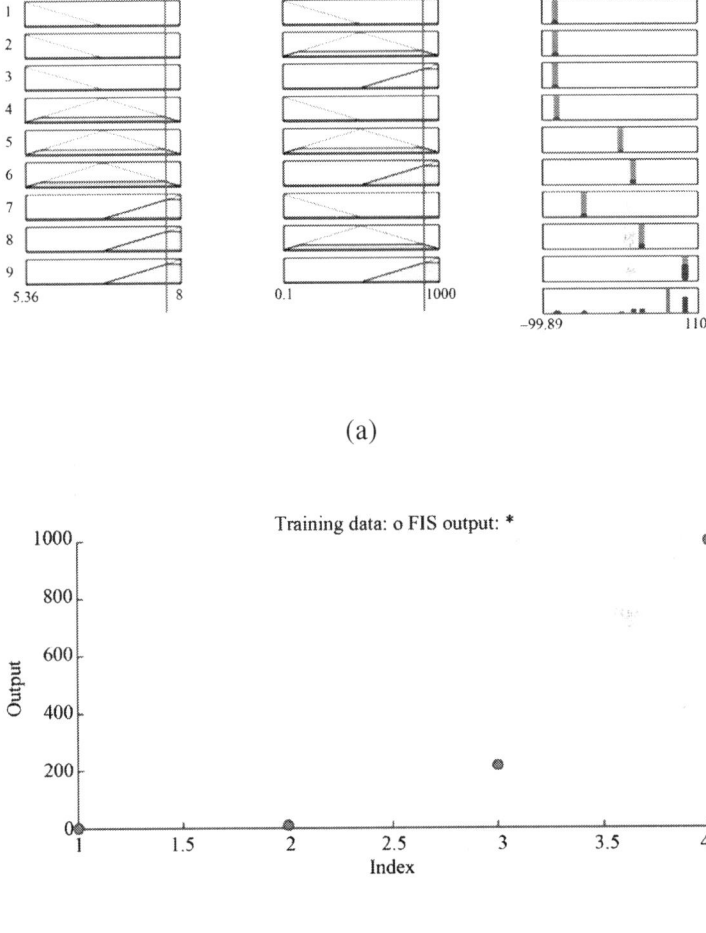

(a)

(b)

Figure 5: (a) ANFIS model and (b) training data: for mercury adsorption to sediments by varying initial Hg concentration and pH.

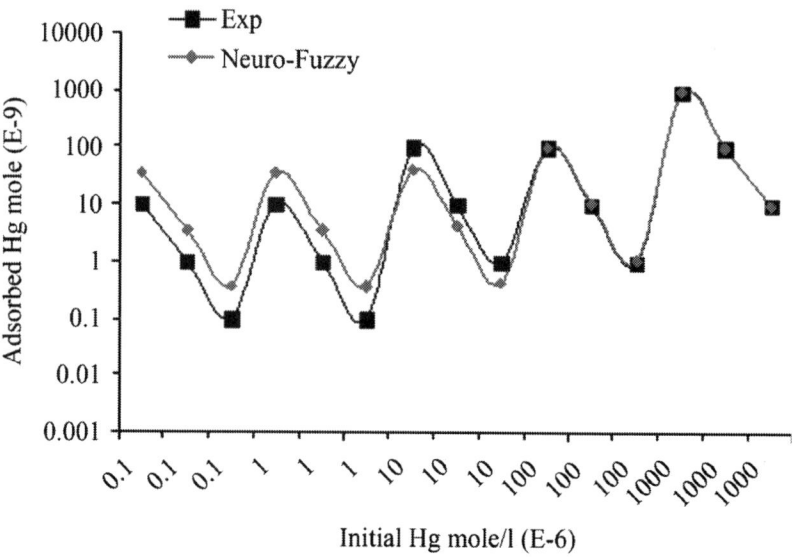

Figure 6: Comparison between ANFIS simulation and experimental data for adsorbed Hg within different initial Hg concentration, when pH is varying as 8, 6.5, and 5.36 respectively at each concentration.

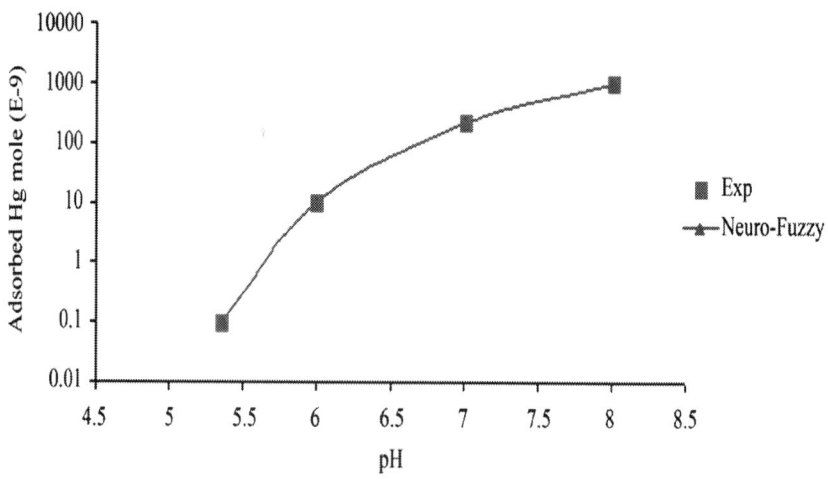

Figure 7: Comparison between ANFIS simulation and experimental data for Adsorbed Hg concentration within variable pH in Solution.

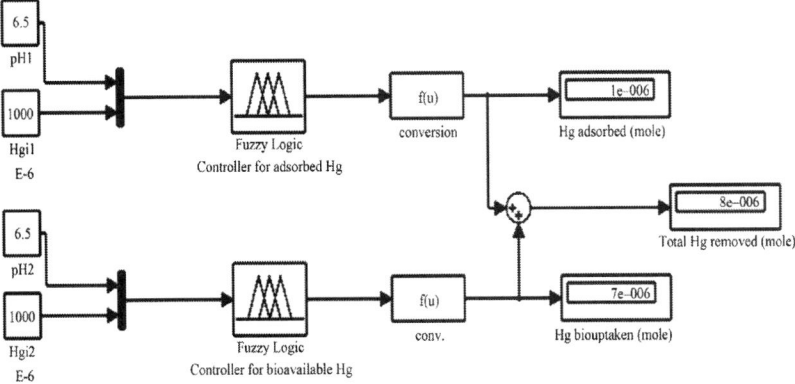

Figure 8: Simulink diagram forecast for total mercury removal for different Hg_i and pH variables.

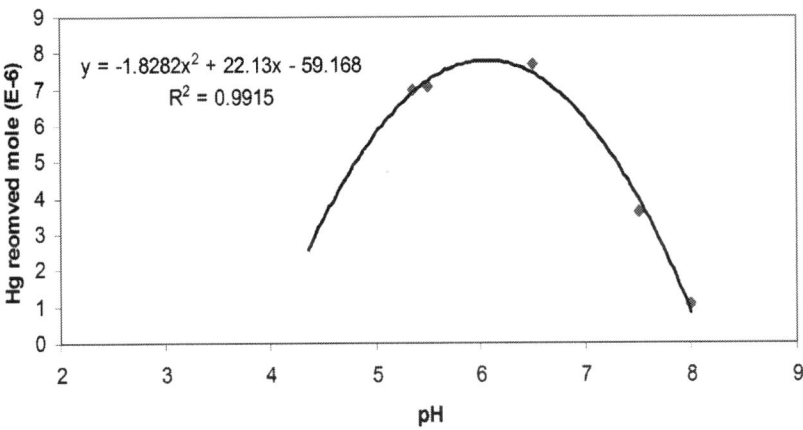

Figure 9: Total mercury removed by natural water within different pH, and the forecast equation.

CONCLUSIONS

Fuzzy logic proved to be useful for assessing ambiguous natural processes. Modeling of mercury bioavailability for bio-species and adsorption to sediments shows strong correlation of more than 98%

between simulation results and experimental data. Using adaptive neuro-fuzzy system is important for hazy system and no experience about data behavior. The findings of this research provide information, simulation, and forecasting about heavy metal removal efficiency in natural systems.

ACKNOWLEDGEMENTS

The financial support from the Natural Sciences and Engineering Research Council of Canada under grant RGPIN-18948 is gratefully acknowledged.

REFERENCES

1.   D. Inthorn, H. Nagase, Y. Isaji, K. Hirata and K. Miyamoto, "Removal of Cadmium from Aqueous Solution by the Filamentous Cyanobacterium Tolypothrix tenuis," Journal of Fermentation and Bioengineering, Vol. 82, No. 6, 1996, pp. 580-584.doi:10.1016/S0922-338X(97)81256-1

2.   L. C. Rai, J. P. Gaur and H. D. Kumar, "Phycology and Heavymetal Pollution," Biological Reviews, Vol. 56, No. 2, 1981, pp. 99-103. doi:10.1111/j.1469-185X.1981.tb00345.x

3.   V. Kuppusamy, J. R. Jegan, K. Palanivelu and M. Velan, "Copper Removal from Aqueous Solution by Marine Green Alga Ulva Reticulate," Electronic Journal of Biotechnology, Vol. 7, No. 1, 2004, pp. 61-71.

4.   M. Horsfall Jr. and A. I. Spiff, "Effects of Temperature on the Sorption of Pb^{2+} and $Cd^{2+}$from Aqueous Solution by Caladium Bicolour (wild Cocoyam) Biomass," Electronic Journal of Biotechnology, Vol. 8, No. 2, 2005.

5.   J. C. Igwe and A. A. Abia, "Maize Cob and Husk as Adsorbents for Removal of Cd, Pb and Zn Ions from Wastewater," The Physical Science, Vol. 2, 2003, pp. 83-94.

6.   L. H. Keith and W. A. Telliard, "Priority Pollutants," Environmental Science & Technology, Vol. 13, No. 4, 1979, pp. 416-424. doi:10.1021/es60152a601

7. B. Volesky and Z. R. Holan, "Biosorption of Heavy Metals," Biotechnology Progress, Vol. 11, No. 3, 1995, pp. 235-250. doi:10.1021/bp00033a001

8. J. L. Wang and C. Chen, "Biosorption of Heavy Metals by Saccharomyces Cerevisiae: A Review," Biotechnology Advances, Vol. 24, No. 5, 2006, pp. 427-451.doi:10.1016/j.biotechadv.2006.03.001

9. K. Vijayaraghavan and Y. S. Yun, "Bacterial Biosorbents and Biosorption," Biotechnology Advances, Vol. 26, No. 2, 2008, pp. 266-291. doi:10.1016/j.biotechadv.2008.02.002

10. G. M. Gadd, "Interactions of Fungi with Toxic Metals," New Phytologist, Vol. 124, No. 1, 1993, pp. 25-60. doi:10.1111/j.1469-8137.1993.tb03796.x

11. V. S. Podgorskii, T. P. Kasatkina and O. G. Lozovaia, "Yeasts—Biosorbents of Heavy Metals," Mikrobiol Z, Vol. 66, 2004, pp. 91-103.

12. D. Kratochvil and B. Volesky, "Advances in the Biosorption of Heavy Metals," Trends in Biotechnology, Vol. 16, No. 7, 1998, pp. 291-300. doi:10.1016/S0167-7799(98)01218-9

13. S. Tunali, A. Cabuk and T. Akar, "Removal of Lead and Copper Ions from Aqueous Solutions by Bacterial Strain Isolated from Soil," Chemical Engineering Journal, Vol. 115, No. 3, 2006, pp. 203-211. doi:10.1016/j.cej.2005.09.023

14. M. X. Loukidou, T. D. Karapantsios, A. I. Zouboulis and K. A. Matis, "Diffusion Kinetic Study of Cadmiurn (II) Biosorption by Aeromonas caviae," Journal of Chemical Technology and Biotechnology, Vol. 79, No. 7, 2004, pp. 711-719. doi:10.1002/jctb.1043

15. J. A. Davis, B. Volesky and R. H. S. F. Vierra, "Sargassum Seaweed as Biosorbent for Heavy Metals," Water Research, Vol. 34, No. 17, 2000, pp. 4270-4278.doi:10.1016/S0043-1354(00)00177-9

16. A. Selatnia, A. Boukazoula, N. Kechid, M. Z. Bakhti and A. Chergui, "Biosorption of Fe^{3+}from Aqueous Solution by a Bacterial Dead Streptomyces Rimosus Biomass," Process Biochemistry, Vol. 39, No. 11, 2004, pp. 1643-1651. doi:10.1016/S0032-9592(03)00305-4

17.   T. Srinath, T. Verma, P. W. Ramteke and S. K. Garg, "Chromium (VI) Biosorption and Bioaccumulation by Chromate Resistant Bacteria," Chemosphere, Vol. 48, No. 4, 2002, pp. 427-435. doi:10.1016/S0045-6535(02)00089-9

18.   K. Vijayaraghavan, J. R. Jegan, K. Palanivelu and M. Velan, "Copper Removal from Aqueous Solution by Marine Green Alga Ulva Reticulate," Electronic Journal of Biotecnology, Vol. 7, No 1, 2004. doi:10.2225/vol7-issue1-fulltext-4

19.   A. Ozturk, "Removal of Nickel from Aqueous Solution by the Bacterium Bacillus Thuringiensis," Journal of Hazardous Materials, Vol. 147, No. 1-2, 2007, pp. 518-523.doi:10.1016/j.jhazmat.2007.01.047

20.   T. R. Muraleadharan, L. Iyengar and C. Venkobachar, "Screening of Tropical Wood-Rotting Mushrooms for Copper Biosoption," Applied and Environmental Microbiology, Vol. 61, No. 9, 1995, pp. 3507-3508.

21.   A. Nakajima and T. Tsuruta, "Competitive biosorption of thorium and uranium by Micrococcus luteus," Journal of Radioanalytical and Nuclear Chemistry, Vol. 260, No. 1, 2004, pp. 13-18. doi:10.1023/B:JRNC.0000027055.16768.1e

22.   S. Gang and S. Weixing, "Sunflower Stalks as Adsorbents for the Removal of Metal Ions from Wastewater," Industrial & Engineering Chemistry Research, Vol. 37, No. 4, 1998, pp. 1324-1328. doi:10.1021/ie970468j

23.   J. Rincon, F. Gonzalez, A. Ballester, M. L. Blazquez and J. A. Munoz, "Biosorption of Heavy Metals by Chemically-Activated Alga Fucus Vesiculosus," Journal of Chemical Technology and Biotechnology, Vol. 80, No. 12, 2005, pp. 1403-1407. doi:10.1002/jctb.1342

24.   N. Kuyicak and B. Volesky, "Biosorption by Fungal Biomass," In: B. Volesky, Ed., Biosorption of Heavy Metals, CRC Press, Florida, 1990, pp. 173-198.

25.   E. Romera, F. Gonzalez, A. Ballester, M. L. Blazquez and J. A. Munoz, "Biosorption with Algae: A Statistical Review," Critical Reviews in Biotechnology, Vol. 26, No. 4, 2006, pp. 223-235. doi:10.1080/07388550600972153

26. J. D. Allison, D. S. Brown and K. J. Novo-Gradac, "MINTEQA2/ PRODEFA2, Geochemical Assessments Model for Environmental Systems: Version 3.0 User's Manual," Computer Sciences Corporation, Environmental Research Laboratory, Athens, GA, 1991.

27. A. El-Agroudy and M. Elektorowicz, "Kinetics of Inorganic Mercury Removal from Surface Water by Water Hyacinths and Reeds," 34 CCSWPR, Burlington, 1999.

28. A. El-Agroudy, "Investigation of Constructed Wetland Capability to Remove Mercury from Contaminated Waters," Ph.D. Thesis, Concordia University, 1999.

29. E. Maria, B. Marek and Q. Ahmad, "Application of the AI to Estimate the Constructed Wetland Response to Heavy Metal Removal," ASCE/CSCE Conference on Environmental Engineering, Niagara Falls, July 2002.

30. R. J. Hunter, "Introduction to Modern Colloid Science," Oxford University Press Inc., New York, 1993.

31. K. H. Tan, "Principles of Soil Chemistry," Marcel Dekker Inc., New York, 1982.

32. R. N. Yong, A. M. O. Mohamed and B. P. Warketin, "Principles of Contaminant Transport in Soils," Development in Geotechnical Engineering, Elsevier, 1992.

Citations

CHAPTER 1

María Victoria Casares, Laura I. de Cabo, Rafael S. Seoane, et al., "Measured Copper Toxicity to Cnesterodon decemmaculatus (Pisces: Poeciliidae) and Predicted by Biotic Ligand Model in Pilcomayo River Water: A Step for a Cross-Fish-Species Extrapolation," Journal of Toxicology, vol. 2012, Article ID 849315, 11 pages, 2012. doi:10.1155/2012/849315.

CHAPTER 2

Javed Iqbal and Munir H Shah, Occurrence, Risk Assessment, and Source Apportionment of Heavy Metals in Surface Sediments from Khanpur Lake, Pakistan, doi:10.1186/s40543-014-0028-z.

CHAPTER 3

Nurdan Gamze Turan, Emine Beril Gümüşel, and Okan Ozgonenel, "Prediction of Heavy Metal Removal by Different Liner Materials from Landfill Leachate: Modeling of Experimental Results Using Artificial Intelligence Technique," The Scientific World Journal, vol. 2013, Article ID 240158, 5 pages, 2013. doi:10.1155/2013/240158.

CHAPTER 4

Pulatsü, S. and Topçu, A. (2015) Review of 15 Years of Research on Sediment Heavy Metal Contents and Sediment Nutrient Release in Inland Aquatic Ecosystems, Turkey. Journal of Water Resource and Protection, 7, 85-100. doi: 10.4236/jwarp.2015.72007.

CHAPTER 5

Stanko Ilić Popov, Trajče Stafilov, Robert Šajn, Claudiu Tănăselia, and Katerina Bačeva, "Applying of Factor Analyses for Determination of Trace Elements Distribution in Water from River Vardar and Its Tributaries, Macedonia/Greece," The Scientific World Journal, vol. 2014, Article ID 809253, 11 pages, 2014. doi:10.1155/2014/809253.

CHAPTER 6

Mabika, N., Masiya, T., Utete, B. , Barson, M. and Tsamba, J. (2015) Trace Metal Concentration in Two Matrices in an Urban Subtropical River. Journal of Water Resource and Protection, 7, 219-227. doi: 10.4236/jwarp.2015.73018.

CHAPTER 7

Robert J. Martinez, Melanie J. Beazley, and Patricia A. Sobecky, "Phosphate-Mediated Remediation of Metals and Radionuclides," Advances in Ecology, vol. 2014, Article ID 786929, 14 pages, 2014. doi:10.1155/2014/786929.

CHAPTER 8

F. Nasehi, A. Hassani, M. Monavvari, A. Karbassi, N. Khorasani and A. Imani, "Heavy Metal Distributions in Water of the Aras River, Ardabil, Iran," Journal of Water Resource and Protection, Vol. 4 No. 2, 2012, pp. 73-78. doi: 10.4236/jwarp.2012.42009.

CHAPTER 9

A. Qasaimeh, M. Abdallah and F. Bani Hani, "Adaptive Neuro-Fuzzy Logic System for Heavy Metal Sorption in Aquatic Environments," Journal of Water Resource and Protection, Vol. 4 No. 5, 2012, pp. 277-284. doi: 10.4236/jwarp.2012.45030.

Index